What Cerebral Blood Flow Tells Us

Iwao Kanno

What Cerebral Blood Flow Tells Us

Measurement and Fluctuation

Springer

Iwao Kanno
Advanced Neuroimaging Center
National Institutes for Quantum Science and Technology
Chiba, Japan

ISBN 978-981-97-6381-8 ISBN 978-981-97-6382-5 (eBook)
https://doi.org/10.1007/978-981-97-6382-5

The translation was done with the help of artificial intelligence (machine translation by the service DeepL.com). A subsequent human revision was done primarily in terms of content.

Translation from the Japanese language edition: "Nouketsuryuryou ha Kataru" by Iwao Kanno,

This Springer imprint is published by the registered company Springer Nature Singapore Pte Ltd.
The registered company address is: 152 Beach Road, #21-01/04 Gateway East, Singapore 189721, Singapore

Preface

Cerebral blood flow supports the working brain. When the brain is at rest, it provides a steady supply of cerebral blood flow to the working area. When you suffer from some disease like stroke or dementia, the cerebral blood flow will reflect the pathological condition. At the same time, cerebral blood flow fluctuates in time in various regions of the brain depending on the local brain activity. The fluctuation reflects the different levels at which the brain is working moment-to-moment in time and space. However, the mechanism responsible for cerebral blood flow fluctuations is not yet clearly understood. When the brain works, neurons use energy, so it is quite natural that the cerebral blood flow increases where it is needed. You may not understand why this is so strange. Even first-class neuroscientists believe that it is natural that cerebral blood flow increases because it is necessary to supply energy. However, the latest research has shown that even if cerebral blood flow increases locally, it is not all for energy consumption in the local area. Therefore, this book explores how cerebral blood flow so eloquently tells us about the brain's function with temporal and spatial precision.

When the author graduated in electrical engineering from Tohoku University in 1970, Japan was in a period of rapid economic growth. Professors of engineering tried to send their students to big companies, but the author ran away from their recommendations to escape from becoming just a cog in the wheel of a company. At that time, the author heard about the establishment of a "brain institute" in Akita, namely, the Akita Brain and Blood Vessel Research Institute (Akita Brain Institute), and he did not hesitate to go there. Then, like a young monk outside the gate of a temple (a fervent acolyte), the author started learning about the brain and measuring cerebral blood flow. A few years later, he was invited by Niels Lassen to study in Copenhagen for 6 months, and since then he has been unable to leave this field. Immediately after the author returned to Japan, it was decided to expand the Akita Brain Institute, and his supervisor made the decision to install a cyclotron and start a PET facility dedicated to stroke. From then on, he had to spend 24 h a day setting up the PET laboratory, but it was an exciting time. In the end, our efforts paid off, and the accuracy of the PET measurements at the Akita Brain Institute was acknowledged worldwide.

This book also describes the methods for measuring cerebral blood flow, which are now often skipped as already well established. At the same time, we will look back at the breakthroughs in theory and technology made by our predecessors since the era of Kety-Schmidt to the present.

Instruction for Readers

This book is intended to be as simple as possible so that the general public may understand the role of cerebral blood flow in supporting the brain. On the other hand, we hope that specialists on the front line of clinical practice in neurology and neurosurgery might also be able to understand the essence of cerebral blood flow measurement methods and the regulatory mechanism of cerebral blood flow. However, please refer to a specialized physiology textbook for more complete information on cerebral blood flow. Some readers may wish to skip the chapters indicated in bold.

For general students who want to learn about the brain front to back
1, 2, 3, 4, **5**, **6**, **7**, 8, 9

Medical students who want to know about the physiology of cerebral blood flow
1, 2, **3**, 4, **5**, **6**, 7, 8, 9

Clinicians who want to know about the principles of cerebral blood flow measurement
1, **2**, **3**, 4, 5, 6, **7**, **8**, 9

General people who want to learn more about the principles of functional brain imaging
1, 2, 3, 4, **5**, **6**, **7**, 8, 9

Middle-aged and elderly people who want to learn more about cerebrovascular diseases
1, 2, **3**, 4, **5**, **6**, 7, **8**, 9

Chiba, Japan Iwao Kanno

Acknowledgment

The author is grateful to his two mentors: Kazuo Uemura, at the Akita Brain Institute, and Niels A. Lassen, at Bispebjerg Hospital. Unfortunately, both have now passed away. The author is also grateful to his colleagues at the Akita Brain Institute and Molecular Imaging Center at the National Institute of Radiological Sciences (NIRS). The author is also grateful to Jeff Kershaw for critically checking the English phrasing and logic while editing the book. Finally, the author thanks his family for allowing him to spend time satisfying the curiosity he has felt ever since he entered this field.

About the Author

Iwao Kanno graduated in Electric Engineering from Tohoku University, Sendai, Japan in 1970. He started work in the Department of Radiology and Nuclear Medicine at the Akita Brain and Blood Vessel Research Institute. He was a visiting researcher at Niels Lassen's laboratory at Bispebjerg Hospital, Copenhagen, Denmark in 1978, and at the MRC Cyclotron Unit at Hammersmith Hospital, London, UK in 1982. He has developed a measurement method for cerebral blood flow and built a custom hybrid PET and SPECT system. He served as Director of the Molecular Imaging Center at the National Institute of Radiological Sciences, Chiba from 2006 to 2010. He is currently a cooperative researcher in the Advanced Neuroimaging Center at the National Institutes for Quantum Science and Technology, Chiba, Japan.

Contents

1 Prologue

Abstract

Currently, all life science uses 3-D imaging. In particular, imaging of brain function at the macroscopic and microscopic scales has developed rapidly. Out of all the various biological elements of the brain, the cerebral blood flow is the most indispensable in supplying oxygen and nutrition. Nevertheless, the cerebral blood flow fluctuates temporally and spatially in tune with brain activity. The biological mechanisms underlying this fluctuation have not yet been clarified. This chapter introduces the author's motivation for writing this monograph.

Keywords

Imaging technology · Fluctuation · Cerebral blood flow · Healthy brain · Aged brain

1.1 Recent Progress in Life Science

Today's life science research is full of imaging. Even in the field of neuroscience, we can now image the morphology and function of the brain from the micro to the macro scale. On the macroscopic level, we can observe the various functions of the whole human brain using magnetic resonance imaging (MRI), positron emission tomography (PET) and single photon emission computed tomography (SPECT). With advances in genetic modification and microscopy technologies, it is possible to express fluorescent proteins as desired in neurons, glial cells, and vascular endothelial cells so that in vivo animal experiments can be conducted to study the flow of red blood cells and plasma in capillaries, changes in blood vessel diameter, calcium ion (Ca^{++}) activity in neurons, and the activity of glial cell processes in real time. From the data, we are now able to measure the microscopic relationship between neural activity and blood flow metabolism in three dimensions deep into the cerebral cortex.

1.2 The Universe According to Cerebral Blood Flow

Cerebral blood flow indicates the state of health and disease of the brain. At rest, cerebral blood flow is distributed to each region according to the density of neurons. Cerebral blood flow of elderly people with cerebrovascular disease, dementia, and other brain disorders changes in the region where the disease is present. Therefore, just by measuring cerebral blood flow, the causative area of diseases can be identified. Cerebral blood flow measurement has become a powerful diagnostic tool for brain diseases and is an indispensable test for the clinical diagnosis of the brain and nervous system.

I. Kanno, *What Cerebral Blood Flow Tells Us*, https://doi.org/10.1007/978-981-97-6382-5_1

However, with the advancement of measurement technology, it has become possible to indirectly estimate cerebral blood flow with MRI over time. This has led to the discovery that cerebral blood flow fluctuates. Moreover, it has become clear that these fluctuations correspond to regions containing active neurons in the brain. When the fluctuations in cerebral blood flow are mapped onto the three-dimensional space of the brain, it is evident that the fluctuations are synchronized with each other in the brain and that the regions form neural networks. Cerebral blood flow fluctuates as if it represents the activity in the brain.

For this reason, some people naively believe that the era of being able to read people's minds by combining brain function mapping information with artificial intelligence (AI) technology is just around the corner. However, at present, the spatial resolution of MRI technology that can map the entire brain is too coarse, at a maximum of 4–5 mm. This resolution is very far from the fineness of the neurons that make up the actual unit of brain function. It may still take some time to collect and interpret all the various research on the individual regions of the brain that has been performed around the world to date, but when analysis of this so-called *big data* becomes possible, the day may come when we can understand the essence of the mind. However, there is one thing that can be said with certainty even today: the mind and spirit are products of the activity of neurons.

1.3 A Long-Awaited Model with New Dimensions

Even though it was said that an AI computer could never beat a professional Go player, in 2018 it won. We are approaching a frightening era in which AI computers will someday have minds of their own, and the possibility of them invading the realm of humans is just around the corner. Brain function mapping information supports the foundation of brain science, which continues its remarkable development, but the source of the signal is just a very simple change in cerebral blood flow. Microscopically, it has been confirmed in animal experiments that these changes in cerebral blood flow reflect neuronal activity. However, it has not yet been confirmed whether those changes, and the corresponding mapping information, faithfully reflect the highly complex neural activities corresponding to higher brain functions in humans. Nevertheless, we are now trying to understand consciousness and perhaps even the deep psychology of the brain. It is a mystery why cerebral blood flow is so fluently and precisely in tune with brain activity. Why neural activity is represented by changes in cerebral blood flow is unknown and will require a more advanced analytical model before an answer can be found.

1.4 Era When the Author Started Measuring Cerebral Blood Flow

In the 1970s, cerebral blood flow provided important information for neurological medicine. At that time, CT and cerebral blood flow could not be measured widely, and the most reliable examination for central nervous system diseases was cerebral angiography. Patients with an abnormality in the central nervous system underwent cerebral angiography first to rule out the possibility of cerebrovascular disease. Based on cerebral angiography, we had to determine whether it was a stroke, brain tumor, or other brain disease, and if it was a stroke, we had to identify the responsible blood vessel, which was a process of trial and error. The neurosurgery, neurology, and radiology departments bickered and argued about the interpretation of the cerebral angiogram. The distribution of cerebral blood flow, as well as the prognosis and treatment for a patient's clinical symptoms, was estimated based solely on the information from cerebral angiography.

Column: Akita Method

Nearly 50 years ago, Prime Minister Eisaku Sato (1901–1975) had a stroke attack at a Japanese restaurant in Tsukiji in Tokyo. The story is well known that, following the instructions of a famous professor at Tokyo University who was in control of the field of internal medicine at that time, Prime Minister Sato was kept at rest for several days in the restaurant without examination. At that time, the Akita Cerebrovascular Research Center (Akita Brain Institute), a young group of clinical specialists for stroke in a rural city in northern Japan, had established an innovative new method. They developed a diagnosis and treatment method for stroke called the "Akita method", in which patients who suffered a stroke were immediately transported to the hospital by ambulance for angiography to diagnose cerebral hemorrhage or cerebral infarction as soon as possible and start the necessary treatment. However, for the authorities at Tokyo University, the stroke treatment method that had been started in the remote northern countryside seemed reckless. The former Prime Minister, Mr. Sato, was finally rushed to the university hospital five days after the stroke onset, but it was already too late, and he died two weeks later. The "Akita method" has now become a standard procedure. It is clear that the sooner the patient is transported to the hospital for the necessary diagnosis and treatment, the higher the lifesaving rate and the fewer aftereffects.

The situation has changed dramatically over the last 50 years. Today, the measurement of cerebral blood flow associated with cerebral ischemia in cerebrovascular diseases is a matter of course, and how it behaves over time is well established. It is a blessed era, or to put it another way from the standpoint of clinical research, there is almost no uncharted territory left regarding cerebral blood flow. For young clinicians in neurology and neurosurgery, cerebral blood flow for disease diagnosis is not taken as a research subject. However, the author hopes that the description of cerebral blood flow presented in this book will help young researchers rediscover what the author believes is the still incompletely known world of cerebral blood flow.

Cerebral Blood Flow: An Important Feature of the Brain

2

Abstract

This chapter describes what cerebral blood flow is. Cerebral blood flow supplies oxygen and nutrients to the brain. The brain consumes approximately 20 times more oxygen than any other part of the body, and therefore it is vulnerable to oxygen shortage. To feed the brain's greed for oxygen is the most important role of cerebral blood flow. The cerebral blood flow does not only supply oxygen and nutrients but also removes waste produced during brain function. The blood vessels and cerebral blood flow have been neglected by neuroscientists as part of the neurological system. Rather, they treat them as infrastructure for the brain to survive. The cerebral blood flow is spatially and temporarily adjusted to the regional demands of the brain. Global and local cerebral blood flow is precisely regulated during neuronal activity even for pathological and aged brains. Neuroscientists only began to recognize the importance of the circulation system after endogenous optical signals were discovered. The vascular system is also very sensitive to traumatic events.

Keywords

Oxygen and nutrition · Brain waste · Diseases and aging · Neuronal activity · Cerebral blood flow · Fluctuation · Infrastructure

2.1 Cerebral Blood Flow Feeds the Brain's Greed for Oxygen

Evolution has provided the mammalian brain with an extraordinary number of neurons. As the brain developed, neurons occupied the surface of the cerebrum at a very high density. However, because the space within the cranium is limited, to extend the surface area of the cerebral cortex the brain formed gyri and sulci structures that burrow deeper and deeper into the brain like wrinkles (Fig. 2.1). The human brain is made up of tens of billions of neurons, and it is said that there are ten times as many glial cells as neurons. To support these brain tissues, capillaries are densely distributed throughout the brain. Capillaries are tightly packed in the brain, and on average, they are spaced approximately 40 to 50 μm apart from each other. If you look at a cast of the cerebral vasculature made by injecting resin into the blood vessels of a dead person's brain, you can see the densely-packed system of blood vessels. It seems to fill up all the space as if the brain were made up of blood vessels alone. The brain is supported by cerebral blood flow. Compared to the other parts of the body, the brain is only 2% of the body's weight, yet it is supplied by 16% of the cardiac output in blood flow and accounts for 20% of the body's total oxygen consumption (Table 2.1).

I. Kanno, *What Cerebral Blood Flow Tells Us*, https://doi.org/10.1007/978-981-97-6382-5_2

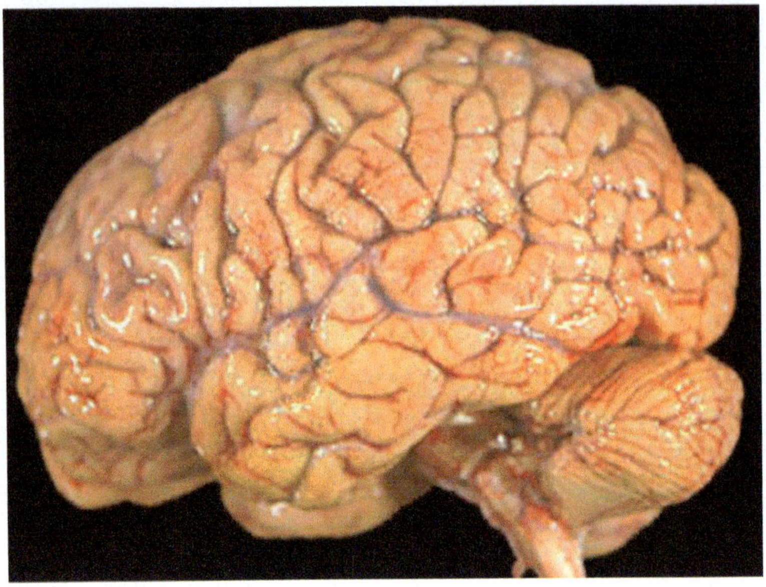

Fig. 2.1 A healthy brain. Left is frontal and right is occipital

Table 2.1 Fraction of Brain to Whole Body

Weight	2%
Cerebral blood flow	16%
Cerebral metabolic rate of oxygen	20%

In the adult brain, cerebral blood flow is on average 50 mL/100 g/min. This means that for every 100 g of brain tissue, 50 mL of blood flows per minute. The resting cerebral blood flow does not change significantly regardless of whether we are asleep or awake. Therefore, as the weight of an average brain is approximately 1500 g, 50 mL/100 g/min × 1500 g = 750 mL/min of blood flows to the whole brain every minute. As a day is 24 hours, 750 mL × 60 min × 24 h ≈ 1000 L (= 1 m^3) of blood is supplied into the brain every day. Furthermore, since the cerebral oxygen consumption is 3.5 mL/100 g/min, the oxygen volume consumed by the brain is 3.5 mL/100 g/min × 150 0 g × 60 m × 24 h = 75.6 L every day. It can be said that the brain is the most gluttonous organ for oxygen. However, if cerebral blood flow is interrupted even for a short period of time, neurons will immediately become dysfunctional, and if it is interrupted for a few minutes, neurons will eventually develop irreversible abnormalities, and necrosis of neurons will begin. Therefore, the most important role of resting cerebral blood flow is to provide a stable supply of oxygen and glucose, collectively called metabolic substrates, as energy sources. The blood flow also receives endocrine hormones and other chemical substances from the nerve nuclei in the brain and other organs and transports them throughout the body via the blood vessels to transmit information.

2.2 Cerebral Blood Flow Cleans Up Brain Waste

As well as supplying nutrients, cerebral blood flow plays the role of a sweeper that carries away the waste that accumulates in the brain. As a new trend in research, there is a hypothesis that toxic substances (e.g., abnormal proteins) that accumulate in brain tissues destroy neurons over time and cause diseases such as dementia, and various studies have reported that cerebral blood flow washes out these abnormal proteins before they become fixed in place (see Sect. 7.4.3). It has been found that people who have suffered traumatic brain injuries and athletes who have received repeated and intense blows to the head, such as boxers, have a higher probability of developing dementia and degenerative brain diseases decades after the injury. Former professional boxer Muhammad Ali, who lit the torch at the opening ceremony of the Atlanta Olympics, had Parkinson's disease. It is thought that Parkinson's disease may be an aftereffect of a head injury, due to the failure of a mechanism that healthy brains have, and which cleans abnormal proteins before they can accumulate in the brain. It has also been recently postulated that lymphatic flow through the brain, which is driven by the beating of the arteries, also cleans the waste that accumulates in the brain (Nedergaard 2013).

Furthermore, the relationship between the brain and sports exercise has attracted attention, and it has been reported that the activation of brain functions by exercise helps prevent brain aging. This is thought to be because exercise actively promotes the mechanism that cleans up waste. Exercise increases local cerebral blood flow, which is thought to promote the discharge of waste in the related area and washes out the waste that accumulates in the surrounding area, but the specific mechanism of the relationship between sports and brain aging prevention has not yet been clarified.

2.3 Remotely Controlled Cerebral Blood Flow

When cerebral blood flow in some region is reduced due to cerebrovascular damage, neurons in the affected area atrophy and die. This results in a phenomenon called remote impairment or secondary neurofunctional dissociation, in which brain function in remote areas is also reduced in conjunction. In this way, a negative chain of events occurs in which an injury decreases brain activity in distant regions and reduces cerebral blood flow to that region. When the middle cerebral artery is infarcted, blood flow and metabolism is reduced on the contralateral side of the cerebellum, which is often seen in stroke patients. This is called "crossed cerebellar diaschisis" (CCD) (Baron et al. 1980). The neural fiber bundles connecting the cerebrum and cerebellum cross from left to right (and vice versa) in the medulla oblongata, and when the middle cerebral artery territory is damaged, the signal to the cerebellum on the opposite side of the cerebrum is cut off, resulting in a decrease in cerebellar blood flow and metabolism.

It is also well known that a chronic decrease in cerebral blood flow around the ventricles leads to a disorder called white matter degeneration, in which the nerve fiber bundles running through the white matter neighboring the ventricles atrophy. It has been suggested that this might result in vascular dementia. Thus, even if the brain tissue in the region and its proximal (upstream) cerebral circulation are normal, or if there is a decrease in brain activity due to damage to the brain region to which the nerve fibers in contact with the region project or damage to the contact path, brain activity will decrease in conjunction with this. Such a remote inhibitory effect has been clearly evaluated only when cerebral blood flow can be measured quantitatively.

2.4 Resting Cerebral Blood Flow Reflects Disease and Aging

Resting cerebral blood flow supplies only what is needed where it is needed. In healthy adults, cerebral blood flow in the resting-state averages 50 mL/100 g/min throughout the brain. If the brain is affected by cerebral ischemia or dementia, cerebral blood flow will be reduced in the regions affected by the disease. Thus, if we know the cerebral blood flow at rest, we can predict the existence of some brain disease. The measurement of resting cerebral blood flow is used in clinical practice as an important diagnostic tool for brain diseases. For example, cerebral blood flow measurement in stroke successfully shows the extent of cerebral ischemia, including the associated degradation of function in distant regions, depending on the area of cerebral blood vessels that have been damaged. Since different types of dementia affect different areas of the brain in different ways, it is possible to estimate the type of dementia to some extent from the pattern of cerebral blood flow. Software for the statistical analysis of cerebral blood flow distribution has been developed for this purpose and is widely applied clinically to support diagnosis.

In healthy adults, the average cerebral blood flow at rest decreases due to the reduction of neurons with aging. In diseased patients, regional

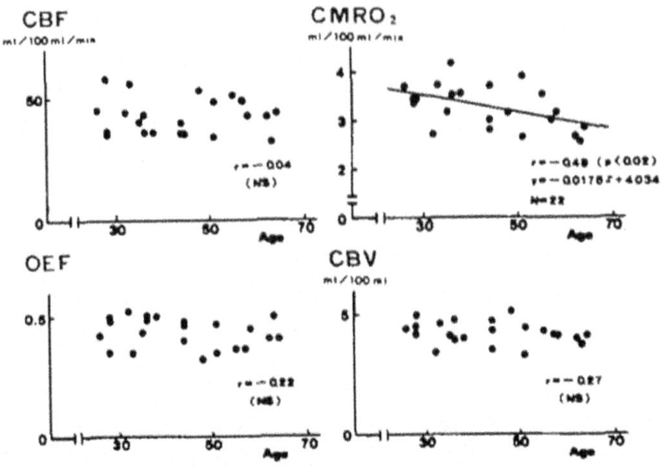

Fig. 2.2 Age dependence of cerebral blood flow (CBF, upper left), cerebral oxygen metabolism (CMRO2, upper right), oxygen extraction fraction (OEF, lower left) and cerebral blood volume (CBV, lower right) measured with PET. CMRO2 is the only quantity that is significantly dependent on age. (Yamaguchi et al. Stroke 1986)

cerebral blood flow may be reduced by the suppression of brain activity due to remote effects related to disease. One early PET study measured cerebral blood flow and cerebral oxygen metabolism in healthy humans with respect to age. The average oxygen metabolism decreased with age after reaching adulthood (Fig. 2.2). Independently, long-term observation of subjects with morphological MR imaging showed that brain volume atrophies with age, and the volume of gray matter, which represents neurons, decreases. This means that the number and density of neurons decreases with aging. When the author was a child, he was told that 200,000 neurons are destroyed every day in the brain. This was used as an excuse for parents to encourage or even threaten children to study harder. It is not clear where the number of 200,000 came from, but it turns out to be quite a trivial amount in comparison to the 100s of billions of neurons that are in the brain.

2.5 Fluctuations in Cerebral Blood Flow Reflect Neural Activity

Cerebral blood flow, on the other hand, changes slightly from time to time, reflecting the functioning of the brain. It is like the brain is muttering to itself. Niels Lassen (Fig. 2.3) developed a method for measuring regional cerebral blood flow using xenon-133. He developed a detector that could measure the distribution of cerebral blood flow in the cerebral hemispheres, and he measured it while giving subjects various tasks to perform successively at intervals of approximately 10 minutes. Since the regional cerebral blood flow increases in response to the tasks given to the subjects, the regions with increased cerebral blood flow are revealed by subtracting the xenon-133 image obtained during the resting

Fig. 2.3 Niels A. Lassen (1926–1997)

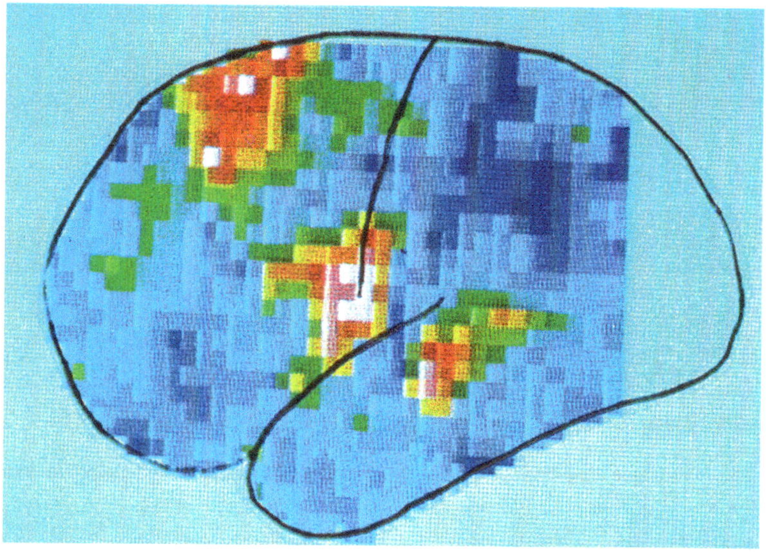

Fig. 2.4 Local increase in cerebral blood flow during an activation task: Cerebral blood flow was measured twice with xenon-133, once during rest and once during conversation. The subtraction of the two cerebral blood flow images of the left hemisphere is shown. The red and yellow areas are foci where cerebral blood flow increases during conversation (Left is frontal, right is occipital). (Lassen NA et al, Sci Am 1978)

state from the xenon-133 image acquired during the task. This was the first time that regional cerebral blood flow changes corresponding to a task were successfully imaged in a healthy person (Fig. 2.4) (Lassen et al. 1978). It was also the first time that we became aware of the brain's murmuring. Although xenon-133 imaging was a coarse method of measuring cerebral blood flow as it had a resolution of only 1 cm, Lassen's work was the first concrete confirmation of Brodmann's functional brain map (Brodmann 1905) (Fig. 2.5). By measuring the regional cerebral blood flow from outside the skull, Lassen succeeded in non-invasively imaging the functional localization of the brain without opening the skull. It is fortunate that the sensory and motor areas of the brain are located on the surface of the cerebral cortex, so the main active areas of the cerebral cortex can be measured by simply placing the detector parallel to the temporal region and measuring the regional cerebral blood flow on the surface of the cerebral hemisphere.

Column: The Genius of Niels A. Lassen

Niels Lassen was a genius. At the age of 31, before he developed the xenon-133 clearance method for measuring regional cerebral blood flow, he published a review paper on cerebral circulation and metabolism which has become a kind of Bible for the field (Lassen 1959). His achievements in the development of techniques for measuring cerebral blood flow are summarized in Chap. 4, but he also made significant contributions to the field of cerebral ischemic pathophysiology with his brilliant ideas and keen insight. He described how, a few days after a stroke, cerebral blood vessels recanalize and cerebral blood flow recovers. The blood flowing after recanalization is a bright red color, which reflects oxygenated hemoglobin. The renewed cerebral blood flow provides a rich supply

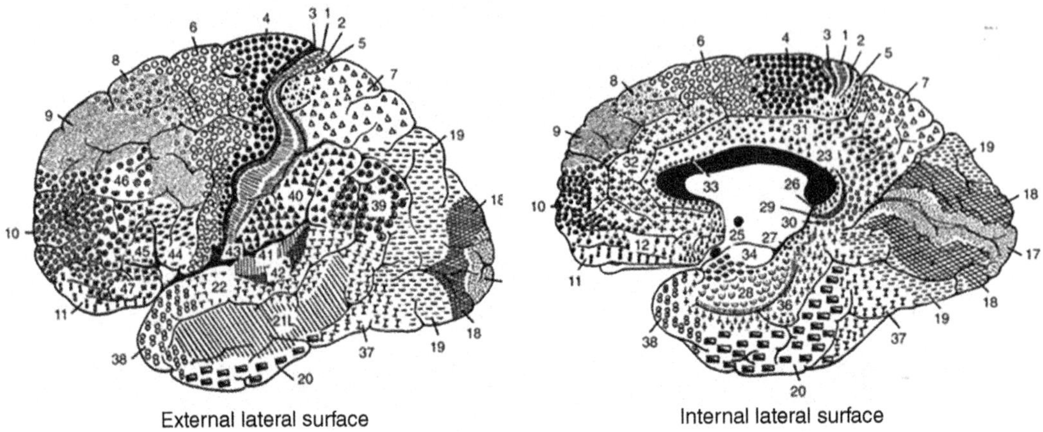

Fig. 2.5 Brodmann map: Brodmann classified brain tissue into 52 regions based on the staining of neurons in the cerebral cortex. (Brodmann KJ, Psvchol Neurol 1905)

of oxygen even though it is not necessary. Hence, he referred to this state as "luxury perfusion". In addition, Lassen clearly described "autoregulation," the now well-known regulation of blood flow in the brain. The concept of the ischemic "penumbra", the border region of necrosis around the ischemic core, was defined by his student Jens Astrup, but the germ of the idea may be embodied in Lassen's work (Astrup et al. 1977).

2.6 Accidental Discovery of Endogenous Optical Signals

Endogenous optical signals that reflect neural activity at the microscale were discovered by Amiram Grinvald in Israel. To observe electrical activity at the cellular level in the cat brain, he put a dye that fluoresces in response to voltage changes onto the brain surface and observed electrical activity in the form of sharp spike-like pulses several milliseconds wide (Fig. 2.6). The spike-like pulses were clearly visible at first, but as the experiment progressed neuronal activity gradually became difficult to observe because of interference from a hazy background light. One day, he synchronized his equipment to the timescale of the slower, hazy light signals and observed them for approximately 50 to 100 milliseconds. To his surprise, the two-dimensional distribution of the signals showed a spatial pattern that was correlated with the neural activity. In an experiment where a striped pattern was shown to a cat, the different responses of the right and left eyes revealed a column structure on the visual cortex (Fig. 2.7) (Grinvald et al. 1986). This was the discovery of an endogenous optical signal. By changing the wavelength of the filter, he found that the signal from reflected white light indicated the difference in oxygenated and deoxygenated hemoglobin in the blood within the visual field. He called this signal an endogenous optical signal as it originated from inside the brain tissue. In contrast to the endogenous optical signal, the fluorescent neuronal signal from the voltage-sensitive dyes, which were his original goal, are called exogenous optical signals because they originate from peripheral nerves outside the brain tissue. Grinvald's experiment was the first time that a neuroscientist, rather than a brain physiologist like Lassen, showed that blood information can be used as a useful signal to measure neural activity.

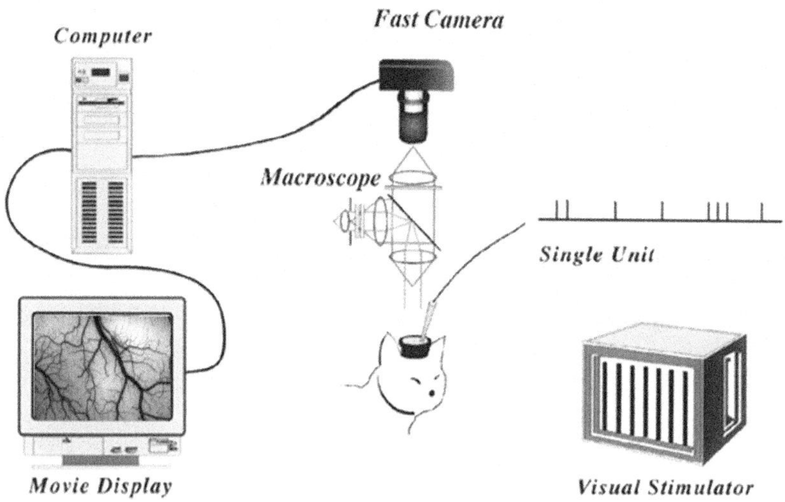

Fig. 2.6 Measurement system for endogenous optical signal: A closed cranial window was prepared on the head of an animal, and the reflected light was measured at two different wavelengths using a high-speed camera microscope during various visual stimulation experiments. The images at the two wavelengths were subtracted to obtain the resulting endogenous signal image. (Reproduced with permission from Grinvald A)

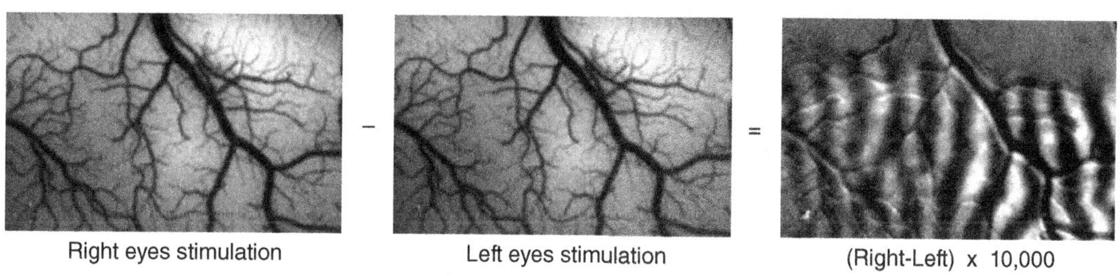

Fig. 2.7 Columnar structure of the primary visual cortex. Intrinsic optical signals were measured during right eye stimulation (left) and during left eye stimulation (middle). After subtracting the two images and scaling the result by 10,000, vertical stripes were revealed (right). (Reproduced with permission from Grinvald A)

2.7 Neuroscientists Considered Cerebral Blood Vessels to Be a Hindrance

Until the discovery of endogenous signals, cerebral blood flow was neglected in neuroscience experiments. Neuroscientists did not consider cerebral blood vessels in their microscopic observations, and the concept of cerebral blood flow probably did not exist for them. Rather, it was a hindrance to the observation of neural activity. When needle electrodes were inserted through the surface of the brain to measure action potentials in animal experiments, damaging blood vessels caused hemorrhaging and prevented the electrode from being positioned as desired. Even if the experiment starts immediately after the animal is awakened from anesthesia after cutting open the skull so that the brain surface can be observed, the nervous system is unaffected if the immediate energy supply is maintained. As far as the damage to the vasculature is local, the nervous system continues to operate normally so that neuroscientists can observe the electrical activity of neurons in their experiments. In general, basic experiments in the field of neuroscience have been acute or short duration experiments that require normal neuronal function to be maintained for only a few hours to half a day to obtain experimental data. In studies that required long-term experiments, such as those that tracked plasticity and learning over days or weeks, the importance of cerebral circulatory metabolism would have been understood, but in short-time neuroscience experiments, the experiment would have been finished before the side effects of abnormal circulation would occur. This suggests that neuroscientists were only aware of the cerebrovascular system as infrastructure for nutrient supply.

2.8 Cerebral Blood Flow Is Hypersensitive to Brain Injury

On the other hand, cerebral ischemia was a major research theme for brain physiologists, and they actively conducted cerebral ischemia experiments on animals. Rodents and small animals were used to study cerebral blood flow regulation and as models of cerebrovascular disease. Cerebral blood flow is very sensitive and vulnerable to experimental injury. To observe the brains of rodents such as rats and mice under a microscope, the skull is opened, and a peephole called a "closed cranial window" is created by cementing a glass plate over the dura mater that covers the brain surface, without hurting it. Even if the surgery to prepare the cranial window is carried out carefully and cautiously, damage to the brain parenchyma is inevitable. If these preparations are not done cautiously in experiments aimed at measuring cerebral circulation and metabolism, the regulatory mechanisms of cerebral circulation and metabolism will soon break down, and correct measurements of cerebral blood flow and metabolism will be disturbed. The vascular system under the dura mater is delicate, and it is well known that surgical operations such as dural incisions can damage this system and break the autoregulation mechanism (see Sect. 7.2.1). To conduct experiments on the vascular system without interfering with the cerebral circulatory and metabolic system, it is necessary to allow time for the brain tissue to adjust to the postoperative environment for approximately a week after the creation of the closed cranial window. Observations of cerebral circulation and metabolism in models of cerebrovascular disorders require stable and reproducible experiments over several weeks to several months, and therefore careful and stable creation of a closed cranial window is essential (Tomita et al. 2005).

Column: Blind Scientists Stroking an Elephant

In the late 1980s, the author had a chance to discuss the brain with some up-and-coming neuroscientists. He learned that the neuroscientists always neglected cerebral blood vessels and cerebral blood flow in their microscopic observations of animal experiments. To neuroscientists, the brain is a factory of "neurons and glia" where neurons are active, and cerebral blood vessels are merely the infrastructure that feeds them. However, to the author, who had been studying cerebrovascular diseases and abnormal cerebral blood flow, the brain was a factory of "cerebral vessels and blood". He was surprised to find such a fundamental difference in viewpoint. The author experienced a similar surprise when he moved from the Akita Cerebrovascular Research Center (Akita Brain Institute) to the National Institute of Radiological Sciences (NIRS). Although both used the same PET technology to observe the brain, NIRS was a factory of "biochemistry" that used PET to research the neurotransmission system. Although they were all focused on the same brain, the neuroscientists, the Akita Brain Institute, and NIRS all had a completely different picture of the brain. They were, and still are, like blind men stroking separate parts of a huge elephant. Each man understands the part that they are touching, but none of them understands the whole elephant.

References

Astrup J, Symon L, et al. Cortical evoked potential and extracellular K+ and H+ at critical levels of brain ischemia. Stroke. 1977;8(1):51–7.

Baron JC, Bousser MG, Comar D, Castaigne P. "Crossed cerebellar diaschisis" in human supratentorial brain infarction. Ann Neurol. 1980;8:128.

Brodmann K. Beitrage zur histologischen localization der Grosshimrinde. Dritte Mitteilung: Die Rinderfelder

niederen Affen. J Phvchol Neurol (Leipzig). 1905;9:177–226.
Grinvald A, Lieke E, et al. Functional architecture of cortex revealed by optical imaging of intrinsic signals. Nature. 1986;324(6095):361–4.
Lassen NA. Cerebral blood flow and oxygen consumption in man. Physiol Rev. 1959;39(2):183–238.
Lassen NA, Ingvar DH, Skinhøj E. Brain function and blood flow. Sci Am. 1978;39(4):62–71.
Nedergaard M. Neuroscience. Garbage truck of the brain. Science. 2013;340(6140):1529–30.
Tomita Y, Kubis N, et al. Long-term in vivo investigation of mouse cerebral microcirculation by fluorescence confocal microscopy in the area of focal ischemia. J Cereb Blood Flow Metab. 2005;25(7):858–67.
Yamaguchi T, Kanno I, et al. Reduction in regional cerebral metabolic rate of oxygen during human aging. Stroke. 1986;17(6):1220–8.

3 Pathways for Cerebral Blood Flow

Abstract

This chapter describes cerebral blood vessel structure, function, and the exchange of gas and nutrients. Vessel anatomy and feeding territories in each of the brain lobes are described. Capillaries, pericytes, glial cells and neurons form an assembly called the neural vascular unit (NVU), which is an interface regulating vascular function. Cerebral blood vessels have a special barrier system, called the blood brain barrier (BBB), to prevent substances moving across the vascular wall. BBB disruption due to some defect may result in cerebral edema, which is excess accumulation of fluid in the intra- or extra-cellular spaces of the brain. O_2/CO_2 and nutrients/waste are supplied/excreted to/from brain tissue through the cerebrovascular network. Finally, the oxygen dissociation from hemoglobin, one of the most important functions of blood flow, is described in detail.

Keywords

Cerebral blood vessel · Vascular network · Blood brain barrier · Oxygen dissociation curve · Hagen-Poiseuille's law · Collateral path

3.1 Structure and Function of Blood Vessels

It is said that the amount of blood in the whole body is approximately 8% of the body weight. For example, a person who weighs 65 kg has a blood volume of approximately 5 liters (L). Most of this blood is circulating. On the other hand, the cardiac output from the heart is said to be 5 L per minute for an adult in a resting state, which means that on average, all the blood passes through the heart once per minute. The blood travels through blood vessels that stretch throughout the body. Starting from the heart, the blood vessels continually branch until they eventually become capillaries. Capillaries then join together again to form venules and veins and finally return to the heart. The total length of the vascular network throughout the body almost reaches 100,000 km, which is two and a half times the circumference of the earth.

3.1.1 Structure of Blood Vessels

The structure of cerebral blood vessels is like those in the body, with arteries made up of three layers: intima, tunica media, and adventitia (Fig. 3.1). The outer membrane is a connective

I. Kanno, *What Cerebral Blood Flow Tells Us*, https://doi.org/10.1007/978-981-97-6382-5_3

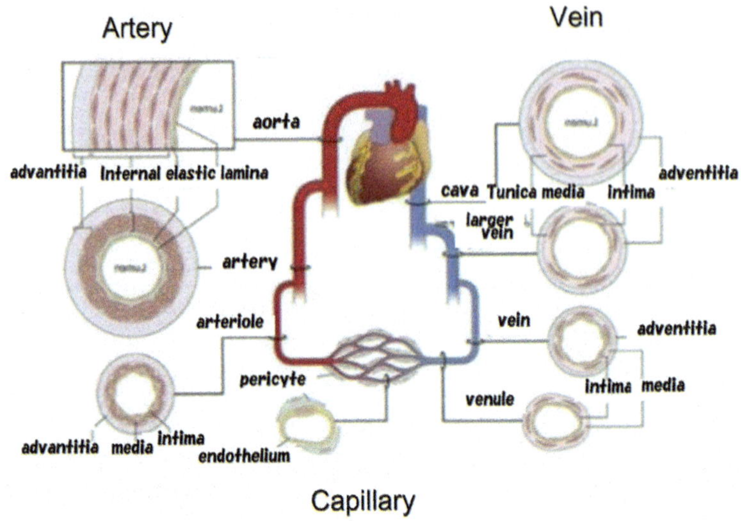

Fig. 3.1 Cross-sectional structure and roles of blood vessels: Arteries, veins, and capillaries have different cross-sectional structures and roles. Arteries regulate blood pressure and cerebral blood flow. Veins act as a reservoir for blood. Capillaries are involved in the exchange of gases and other substances between blood and tissue

tissue that binds to the surrounding tissue. The tunica media is composed of smooth muscle and elastic fibers, and the diameter of blood vessels can be changed by tension and relaxation of the smooth muscle. The intima consists of endothelial cells, inner elastic plate and connective tissue, and regulates the permeation of substances. Veins are also basically made up of three layers of membrane. Arteries have well-developed elastic fibers in the tunica media, while veins have fewer muscle fibers due to low intravascular pressure. Therefore, arteries, which are rich in smooth muscle, are also called elastic vessels. On the other hand, veins store the amount of blood flowing into them and passively expand their diameter to temporarily store blood; therefore, veins are sometimes called capacitance vessels. Approximately 75% of the circulating blood volume is on the venous side. Capillaries are composed of endothelial cells and pericytes. Capillaries exchange gas and nutrients with brain tissue, thus they are called exchange vessels. Endothelial cells play an important role in this process.

3.1.2 Function of Blood Vessels

Blood is pumped out of the left ventricle into the aorta by the heart's pumping function. After this, it flows through the arterial lumen from the proximal side (upstream) to the distal side (downstream). The pressure of arterial blood ejected from the heart pulses due to the beating of the heart, and the maximum and minimum blood pressures oscillate. At the time of ejection, the pressure on the arterial wall (blood pressure) is a maximum of 120–130 mmHg and a minimum of 80–90 mmHg. As the arterial diameter decreases further away from the heart, the amplitude of the pulse wave decreases, and the mean blood pressure decreases. The overall volume of the vascular bed is largest in the capillary region but, conversely, the blood flow velocity is slowest in the capillary region (Fig. 3.2). The arterial vessel wall is elastic and flexible, and the force between the blood and the vessel wall pushes the blood from upstream to downstream. This driving force is called perfusion pressure. When a viscous fluid such as blood flows in a regular laminar fashion,

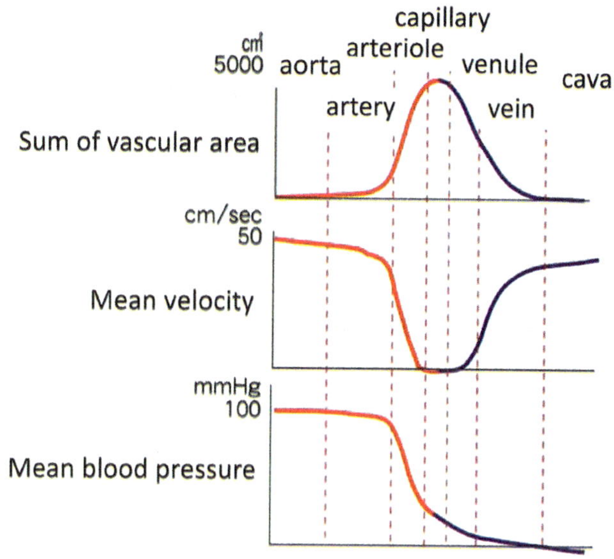

Fig. 3.2 Total area of the vascular cross-section, blood flow velocity and mean blood pressure: The x-axis indicates the path the blood takes, starting from the left ventricle of the heart and ending at the right atrium of the heart. (In Physiology and Biophysics, Elsevier)

the vascular resistance of the blood flowing through the vessel depends on the vessel diameter and is inversely proportional to the fourth power of the vessel diameter according to the Hagen-Poiseuille's law (Fig. 3.3). A slight change in the diameter of the blood vessel will greatly affect vascular resistance.

As the arterial diameter decreases after branching from the aorta, the arterial resistance increases. When the arterial diameter branches to small arteries of approximately 100 μm, the arterial resistance suddenly increases, and the perfusion pressure rapidly drops. After passing through the arterioles, the average blood pressure drops to 30–40 mmHg since the arteriole diameter is even smaller. The smooth muscle of the arterioles is regulated by the sympathetic nervous system, and sympathetic nerve excitation causes the transmitter noradrenaline to act on the smooth muscle, causing it to contract and increase vascular resistance. Since noradrenaline also increases cardiac output, both these effects cause blood pressure to rise. When the arterioles branch and reach the capillaries, the perfusion pressure is thought to be similar to the perfusion pressure in the capillaries of other organs (approximately 4 mmHg). After passing through the capillaries, the blood converges in the venules, and thereafter further repeated merging occurs before reaching the large veins. The blood pressure in the veins is approximately 10 mmHg (Table 3.1). The veins have valves that prevent the backward flow of blood. In the lower limbs and other parts of the body, the venous valves act with muscle movement so that blood returns to the heart against gravity.

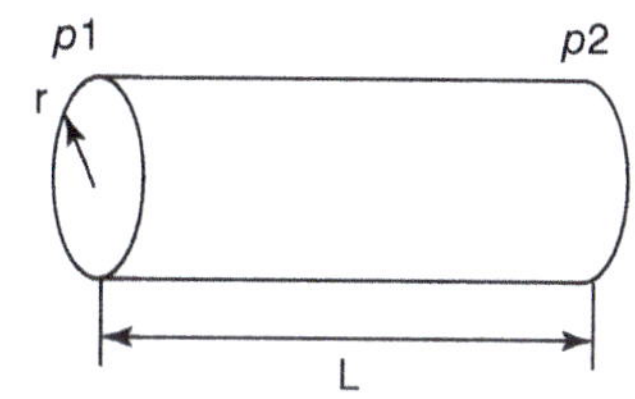

$$\text{Flow rate} = \frac{\pi r^4}{8\mu}\left(\frac{p1 - p2}{L}\right)$$

Fig. 3.3 Hagen-Poiseuille's law: The flow rate of a viscous liquid through a tube is proportional to the fourth power of the tube's diameter (r). μ is the viscosity. p1 and p2 are the blood pressures at the proximal end (1) and distal end (2), respectively, of the vessel of interest

Table 3.1 Vessel diameter, blood pressure and pO_2

Vessel	Artery	Arteriole	Capillary	Venule	Vein
Diameter (μ)	>100	50	5–10	50	>100
BP (mmHg)	80–100	40–80	4–40	4–10	10
pO_2 (mmHg)	100	40	20	?	?
Role	Resistance		Exchange	Capacitance	

3.2 Cerebral Blood Vessels and Surrounding Tissues

The blood vessels in the brain are called cerebral blood vessels, and the blood flow to the brain is called cerebral blood flow. The most important role of cerebral blood flow is to supply oxygen and nutrients to the brain tissue and to discharge waste byproducts produced in the brain tissue. Cerebral blood flow differs from the blood flow environment of other organs in that it flows through a closed space inside the skull. Therefore, it is protected from the pressure of the atmosphere, but it is sensitive to the blood pressure on the blood vessels and the intracranial volume changes of the tissues in the brain. In short, the cerebrovascular network is in a closed space.

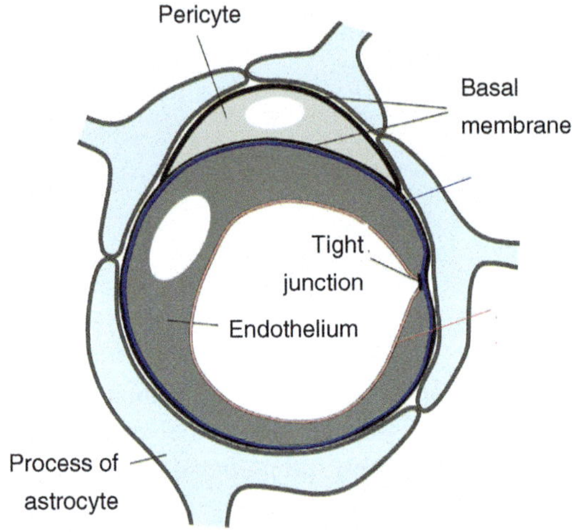

Fig. 3.4 Blood-brain barrier (BBB): Cerebral blood vessels have no structural gaps, and endothelial cells and astrocyte processes wrap around the blood vessels. Pericytes prevent substances from freely moving in and out of the blood vessels

3.2.1 The Blood Brain Barrier (BBB) Protects the Brain

Cerebral blood vessels have an important structure and function different from blood vessels of the other organs. Cerebral blood vessels have a barrier-like structure and function called the blood brain barrier (BBB), which prevents substances in the blood from freely entering and leaving the brain tissue. The walls of ordinary blood vessels have physical gaps that allow relatively free movement of substances in and out of the blood vessels, but the blood vessels of the brain have no physical gaps, and the inner walls of small arteries, capillaries and veins are covered with endothelial cells, while the outer walls are densely covered with astrocyte processes (Fig. 3.4). The astrocyte processes and the endothelial cells lining the inner walls of the blood vessels are mainly responsible for BBB function. Substrates such as glucose, which are necessary for the brain, have their own transporters (or carriers) that actively help to move the substrates to brain tissue. There are also transporters that selectively excrete waste. Gases and small molecules can physically diffuse and pass through to brain tissue. For example, the gases xenon-133 and N_2O (laughing gas), which are tracers that are used for measuring cerebral blood flow, are small in molecular size, chemically inert, and can pass through the BBB freely by physical diffusion. In addition, since the cell membrane of endothelial cells is made of lipid, alcohol and substances that easily dissolve in fat (fat-soluble or lipophilic) can pass through the cell membrane and enter the brain relatively freely. The reason why alcohol and narcotics can freely enter the brain is because they are highly lipophilic and can easily penetrate the cell membrane and enter and leave the brain tissue relatively freely, so the BBB does not function as a barrier to them. They can enter the brain tissue and cause direct effects on neurons.

The important role of the BBB is to prevent the penetration of substances that are not needed in the brain. For example, iodine contrast media used in CT and gadolinium contrast media used in MRI have a large molecular weight and are not necessary for brain tissue, so they cannot pass through the BBB in the normal brain. However, when the BBB fails due to cerebral infarction or brain tumor infiltration, the contrast agent may penetrate the vessel wall and leak into the surrounding tissue. In this way, pathological abnormal areas where BBB function is disrupted can be depicted. This is clinically useful for detecting diseased areas where BBB function is broken due to cerebral infarction or brain tumors.

3.2.2 BBB Disruption and Brain Edema

The BBB is a bulwark for the brain tissue, but when it is damaged by various lesions, such as brain tumors, cerebrovascular disorders, head trauma, and infections, abnormal water accumulates in the brain tissue, and brain edema occurs. There are two types of brain edema: vasogenic edema, in which water accumulates in the extracellular space, and cytotoxic edema, in which water accumulates inside the cells. In the former type, vasogenic edema, water accumulates in the extracellular space mainly due to increased permeability of the vessel wall and leakage of serum proteins. The latter type, cytotoxic edema, is caused by metabolic abnormalities that disrupt the flow of ions in and out of cell membranes, resulting in water accumulation mainly in the cells (Klatzo et al. 1958). Depending on the pathology, cerebral edema can be local or global, resulting in either local or global intracranial pressure increases, respectively. In the case of regional cerebral edema, the intracranial pressure (ICP) in the surrounding area rises, compressing blood vessels and reducing perfusion pressure, resulting in a localized decrease in cerebral blood flow. In the case of diffuse cerebral edema, brain damage is more severe because damage occurs throughout the entire brain, but due to volume expansion in the closed space of the skull, the diffuse intracranial pressure increases, and the pressure concentrates on the vital skull base as a pressure relief route, resulting in a very dangerous situation where cerebral herniation can occur in the worst case. Even in less severe cases, cerebral edema causes an increase in intracranial pressure (ICP), which compresses cerebral blood vessels and reduces cerebral perfusion pressure (CPP), resulting in diffuse hypoperfusion.

Aquaporins (AQPs) were discovered in 1992 as proteins involved in the movement of water into cells, and several subtypes of AQPs in the brain were shown to have an abundance of the subtype aquaporin 4 (AQP4) in astrocyte processes. AQP4 is involved in brain edema caused by BBB failure, especially in cerebral ischemia. Experiments on AQP4 knockout mice show less brain edema due to reduced intracellular water retention. It is not yet known to what extent AQP4 is involved in the movement of water molecules from capillaries to brain tissue. For example, we do not know to what extent AQP4 is involved in the behavior of the tracer oxygen-15-labeled water that is used to measure cerebral blood flow in PET. We need to confirm the effect of AQP4 on oxygen-15 water PET measurements using AQP4 knockout animals.

3.2.3 The Neurovascular Unit (NVU) Interfaces Cerebral Blood Vessels with Neurons

The neurovascular unit (NVU) consists of neurons, cerebral blood vessels, and astrocytes and is considered to be the smallest unit that regulates cerebral blood flow in response to neural activity. Astrocytes are glial cells and have numerous amoeboid processes extending from the cell body (soma); some of the processes touch the cerebral blood vessels while other processes touch the neurons. The NVU plays a central role in neurovascular coupling, which will be discussed later. Recent studies have shown that vascular smooth

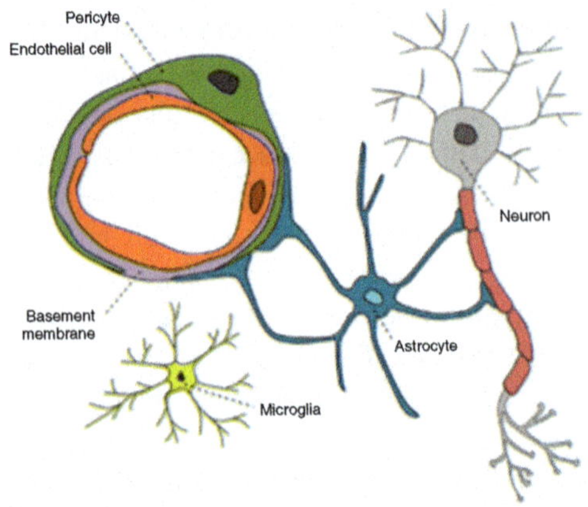

Fig. 3.5 Neurovascular unit (NVU): A neurovascular unit (also sometimes called a neuroglial vascular unit) is composed of endothelial cells, pericytes, neurons, and glial cells such as astrocytes and microglia. (Sweeney MD et al, J Cereb Blood Flow Metab 2015)

muscle, endothelial cells, pericytes, astrocytes, and neurons, all of which play different roles, work together to form the NVU (Fig. 3.5). Cations expressed by astrocytes trigger contraction and relaxation of cerebral vascular smooth muscle, and Ca ions released by neurons also regulate the local cerebral blood flow.

The function of the NVU has been an important subject of research in terms of the regulation of neural activity and cerebral blood flow. However, the full role of NVUs in neurovascular coupling (NVC), where cerebral blood flow increases during neural activation, is still unknown. Advances in genetic engineering have led to the development of genetically modified animals that express fluorescent proteins in a variety of cells, e.g., glial cells, vascular smooth muscle, endothelium, and pericytes. In addition, the mechanism regulating the various functions of the NVU is being investigated by perturbing the NVU through stimulation of the peripheral nerves in the trunk or administration of pharmacologically active substances and observing the activities and changes of individual cells in the NVU in response, acutely or chronically. For example, measuring changes in cerebral vessel diameter over time allows us to track temporal and spatial changes in NVU components in response to peripheral nerve stimulation. It is also possible to visualize Ca ions released during astrocyte stimulation. The roles of the NVU are studied in both healthy and pathologic states. The mechanisms behind neurovascular coupling are summarized in an excellent review (Iadecola 2017).

3.2.4 Glial Cells Support Neurons

Glial cells fill most of the space in the brain. Glia means glue in Greek and refers to the tissue that attaches neurons together. As the name implies, glial cells were thought to be the glue that connects neurons together. Glia includes oligodendrocytes, astrocytes, and microglia, and the total number of glia is thought to be more than ten times that of neurons. Oligodendrocytes wrap around the axons from neurons to form myelin sheaths and provide nutrients to the axons. Astrocytes are the most numerous glial cells and are densely packed in the gray matter. Morphologically, the astrocyte cell body produces a large number of processes, many of which cover the synapses of the neuron body and are responsible for collecting neurotransmitters and regulating the concentration of K^+ and Ca^{2+} cations to assist in the transmission of neuronal signals. Astrocytes extend their projections to neighboring neurons. In addition, astrocytes extend their processes to cover the periphery of neighboring capillaries and arterioles. So, astrocytes act as an interface between neurons and cerebral blood vessels, causing cerebral blood vessels to dilate or contract during neuronal activity, although the precise mechanism is still under debate. Astrocytes also play a central role in the function of the BBB described above, and their glutamine-glutamate shuttle function, in which they take up glucose to produce lactate to provide energy for neurons, has also been well studied (Dienel 2013).

Like macrophages, microglia perform phagocytosis and are activated to repair lesions when inflammation or degeneration occur. It has been shown that they may have a role as immune cells in the central nervous system. The author's group has also observed that they are involved in the sprouting of capillaries (Masamoto et al.

2014), so they have a wide range of roles. The full extent of the role of microglia remains to be elucidated.

3.2.5 Pericytes Assist the Microvessels

Recently, pericytes have come to be considered a component of the NVU. Pericytes themselves have been known to exist since the 1870s, but their functions have only recently been investigated in detail. Pericytes are single cell bodies that surround blood vessels from the arterioles to the capillaries. They are thought to have BBB functions as well as smooth muscle functions that help regulate blood flow through capillaries. Recent imaging with two-photon excitation microscopy has shown that, as well as mid-capillary pericytes, pericytes at the boundary between the arterioles and capillaries actively dilate capillaries to regulate blood flow during neurovascular coupling (NVC) (Attwell et al. 2016; Iadecola 2017). It has also been reported that the so-called "no-reflow phenomenon" (see Sect. 7.4.1), in which capillaries are occluded and local cerebral blood flow is interrupted, occurs because pericytes die while contracting. This phenomenon occurs after transient ischemia because, even if macroscopic cerebral blood flow in large vessels is restored, some microscopic capillaries remain occluded and cannot be restored, leading to the loss of the ability to maintain the BBB and the death of neighboring regions (Dalkara et al. 2011). Thus, by wrapping around capillaries and contracting and expanding them, pericytes have an important function in regulating blood flow at the capillary level (Fig. 3.6) (Hartmann et al. 2015).

3.3 Roles of Cerebral Arteries and Cerebral Veins

This section explains the role of cerebral blood vessels that supply blood into the cranium. Since the brain is an important organ, cerebral blood vessels are characterized by several levels of safety mechanisms against blood flow disruption caused by vascular occlusion and stenosis. These safety mechanisms include the development of collateral blood flow paths such as the circle of Willis arteries in the case of damage to the carotid arteries, cerebral pia mater artery anastomosis in the case of damage to the intracranial trunk cerebral arteries, autoregulation mechanisms in response to blood pressure fluctuations, and redistribution of blood flow in small arteries in the case of damage to peripheral blood vessels. These safety mechanisms will be discussed below.

3.3.1 Cerebrovascular Network

When contraction of the left ventricle of the heart pushes blood out and the intravascular pressure rises, the elasticity of the arterial wall helps to push the blood downstream. When the left ventricle of the heart expands, the intravascular pressure decreases, resulting in pulsatile flow. The relative location of blood vessels may be defined according to the direction of blood flow, with upstream vessels closer to the heart being called "proximal" and downstream vessels farther from the heart being called "distal". From proximal to distal, the aorta repeatedly branches until it reaches the small arteries (approximately 100 μm in diameter). Immediately after the heart beats, there is a large pressure pulse, but the pressure pulse is smoothed

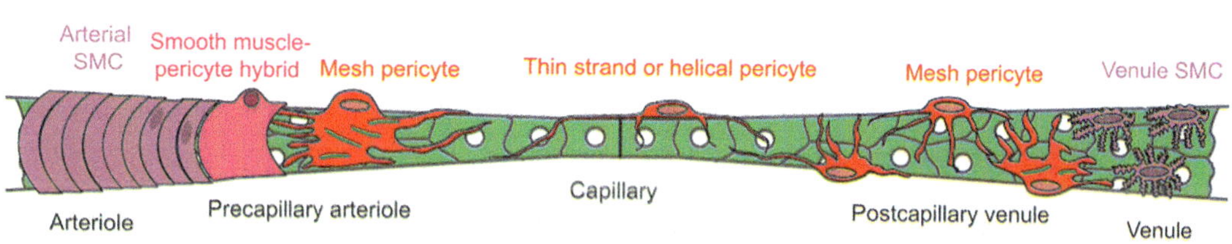

Fig. 3.6 Pericytes: Pericytes of various shapes are found around capillaries across the precapillary, capillary, and post-capillary sections. (Hartmann DA et al. Neurophotonics 2015)

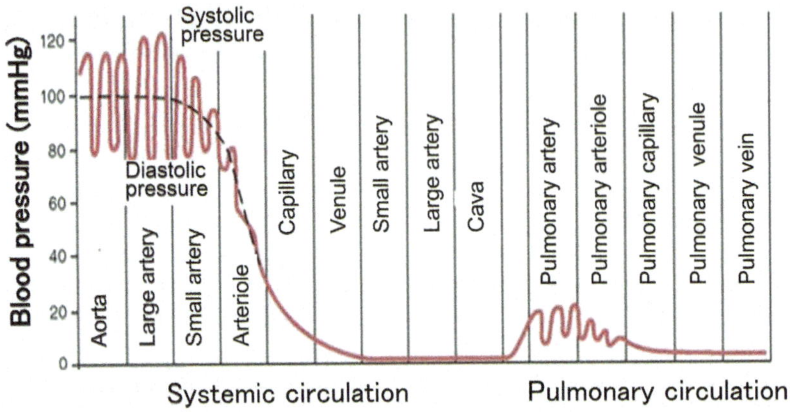

Fig. 3.7 Pulsatile blood pressure for systemic and pulmonary circulations: The pulse pressure is high immediately after leaving the heart. The pressure changes as the blood flows from the body to the pulmonary circulation. The pulse pressure is smoothed out as the vessel diameter decreases. The horizontal axis corresponds to the path the blood takes after leaving the heart and then returning to it. (Medical Physiology ed by Guyton and Hall, Elsevier 2011)

out as it moves toward the distal vessels (Fig. 3.7). The mean arterial pressure in the arteries is approximately 80–100 mmHg. This pressure drops significantly in small arteries with diameter at around the 100 μm level. In other words, the increased vascular resistance due to the narrowed arterial diameter acts as a brake to reduce blood flow velocity, so small arteries are also called resistance vessels. Arterioles have well-developed smooth muscle fibers in the tunica media, and under the control of sympathetic and parasympathetic nerves, the diameter of blood vessels is actively regulated by contraction and dilation. According to the Hagen-Poiseuille's law, a slight change in arterial diameter greatly increases or decreases the arterial resistance of small arteries, thereby regulating blood flow. The capillaries (5–10 μm in diameter) are finally reached after repeated branching from the small arteries. Capillaries are densely distributed in the tissues of organs and exchange substances between the bloodstream and brain tissue and are also called exchange vessels. After passing through the capillaries, the blood flow merges again and flows into the veins. The veins continue to merge and eventually become the vena cava, which returns to the right atrium of the heart. Since most of the blood volume resides on the venous side, the veins are also called capacitance vessels.

3.3.2 The Role of the Major Cerebral Arteries and Veins

The blood vessels that supply blood from the trunk to the head and neck include the carotid artery system, which runs along the anterior aspect of the neck, and the vertebral artery system, which runs along the posterior aspect of the neck. The blood flows into the skull through a total of four arteries, two on each side. The left and right vertebral arteries join as the basilar artery and enter the skull. The intracranial arteries form a ring of blood vessels called the circle of Willis artery, which in turn branches into the middle cerebral artery and the anterior cerebral artery to supply the left and right cerebral hemispheres, and into the posterior cerebral artery to supply the occipital region (Fig. 3.8). The middle cerebral artery is the cerebral artery with the largest perfusion volume, and it supplies cerebral blood flow to the major regions of the cerebral hemispheres as it branches along the lateral surface of the hemispheres. The anterior cerebral arteries are connected by the anterior communicating artery and supply cerebral blood flow mainly to the anterior medial surface of the cerebral hemispheres. The territories within the brain supplied by each of the main trunk cerebral arteries extending from the circle of Willis artery have been determined (Fig. 3.9). However, perfusion pressure inevitably decreases in the border region between the territories corresponding to each main trunk cerebral artery. The border regions are called watershed regions and are regions prone to cerebral ischemia in situations where perfusion pressure decreases, such as when blood pressure drops. In addition, the anterior and posterior communicating arteries that connect the circle of Willis arteries branch directly to the brainstem and important regions of the basal ganglia. The cerebral cortex is divided into four lobes, the frontal, temporal, parietal, and occipital lobes, separated by the main sulci: the central sulcus, lateral sulcus, parieto-occipital sulcus, and anterior occipital sulcus (Fig. 3.10).

The blood in each of the main trunk cerebral arteries branches within the region it supplies to

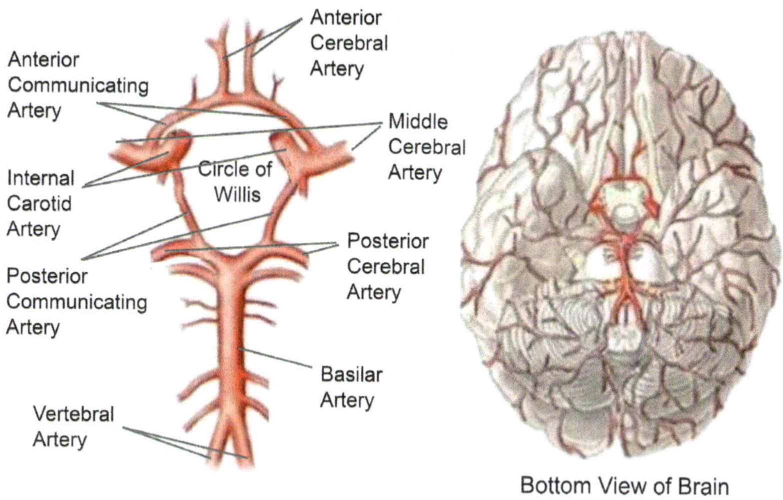

Fig. 3.8 Bottom view of the main arteries in the brain: The left and right internal carotid arteries and the left and right vertebral arteries supply blood to the brain. The left and right middle cerebral arteries, anterior cerebral arteries, and posterior cerebral arteries distribute blood into the various cerebral areas via the Circle of Willis

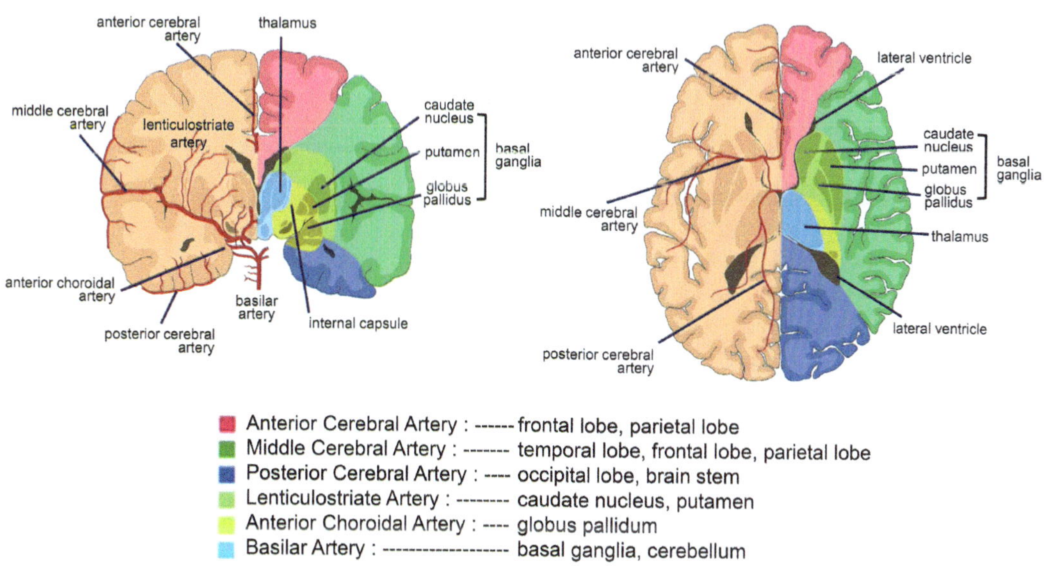

Fig. 3.9 Territories in the cerebrum supplied by each of the cerebral arteries: Each territory is color-coded based on the artery that dominantly supplies that area. (LAIMAN Stockweb Tokyo)

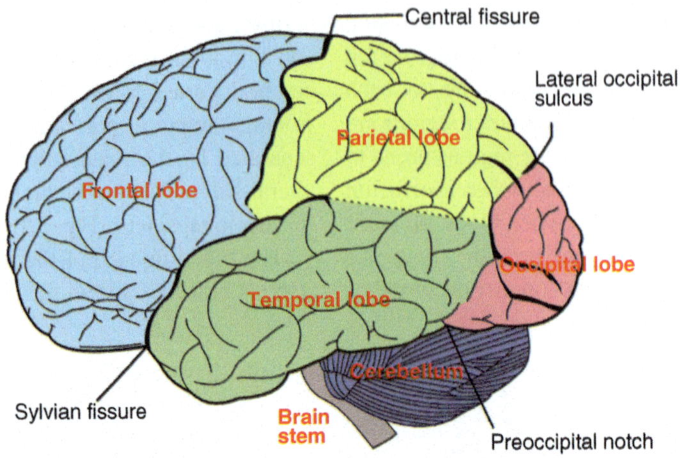

Fig. 3.10 Cerebral lobes: The cerebrum is divided into frontal, parietal, temporal and occipital lobes, which are bordered by the major cerebral sulci (Henry Gray (1918) Anatomy of the Human Body)

eventually perfuse the capillaries, and then returns to the surface of the cerebral cortex through the perforating veins. The veins eventually merge into the anterior cerebral vein, middle cerebral vein, and posterior cerebral vein. The dural venous sinus is a space between the dura mater and the skull that acts as a vein where the confluent venous blood collects. The superior and inferior ends of the sigmoid sinus running through the parietal region are connected to the superior and inferior sagittal venous sinuses, respectively, and the cavernous sinus is located where they meet in the brainstem. Finally, blood flows into the internal jugular vein via the transverse and sigmoid sinuses in the occipital bone.

When discussing the orientation of the brain the following terminology is often used. A cross section that is horizontal to the ground in an upright position is defined as a *horizontal section*. A perpendicular section to the face is defined as a *sagittal section*. This word is used because it is in the direction that an arrow travels when shot from a bow at a person, and its use is said to be based on the legend of William Tell when he shot an arrow at an apple sitting on his son's head. A cross section orthogonal to both the horizontal and sagittal sections is defined as a coronal section because it resembles a crown. The definitions of these cross sections of the brain are shown below (Fig. 3.11).

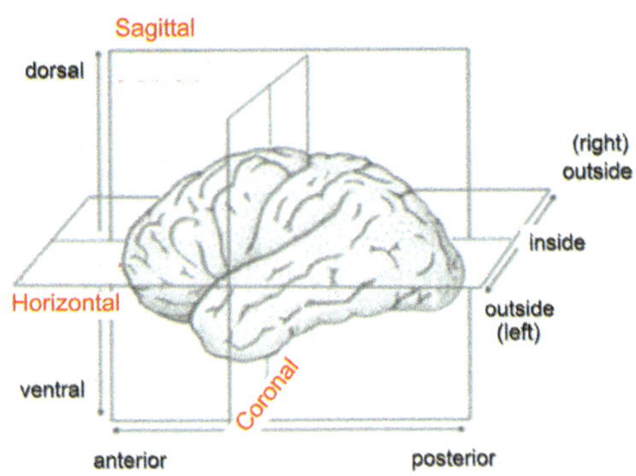

Fig. 3.11 Definitions for the orientation of an anatomical cross-sections of the brain: The plane perpendicular to the anterior-posterior axis is defined as the coronal cross-section, the plane perpendicular to the left-right axis is defined as the sagittal cross-section, and the plane perpendicular to the vertical axis is defined as the horizontal cross-section

3.3.3 The Role of Cerebral Surface Arteries and Perforating Arteries

The major cerebral arteries described in the previous section travel along the surface of the cerebral cortex in a branching fashion. The arteries cover not only the outer surface of the cerebral cortex but also penetrate the inner volume and the folds of the cerebral cortex in the sulcus and gyrus. The arteries that run along the surface of the cerebral cortex are collectively called the cerebral surface arteries or the pial arteries since they run along the pia mater that surrounds the brain. It is thought that the cerebral surface arteries distribute evenly across the cerebral cortex during the development period when the cortex is being formed. Perforating arteries branch out orthogonally from the surface arteries and perforate the brain parenchyma like a plant growing roots into the ground. The intervals between perforating arteries develop in response to the demand for blood flow within the tissue. The perforating arteries branch into capillaries through a series of bifurcations within the brain parenchyma (see also Fig. 6.19). The capillaries then merge and follow the same route in the opposite direction as the perforating artery, ascending to the surface of the brain as the perforating vein, and then merging to form the cerebral surface vein.

3.3.4 Collateral Blood Flow Paths Triggered by an Emergency

The arteries support the brain so that cerebral blood flow on the distal (downstream) side will not decrease even if some arteries are obstructed by occlusion or stenosis. However, in the unlikely event of an obstruction such as stenosis or occlusion of a large artery, nearby dormant blood vessels open to replenish cerebral blood flow. These emergency blood vessels are called collateral blood vessels. Collateral blood supply occurs through three main routes. One frequent route is via the ophthalmic artery and the shallow temporal artery, which connect the intracranial and extracranial arteries. The second route is via the circle of Willis arteries, which is the most important structure in cerebral vascular architecture. The left and right internal carotid arteries and vertebral arteries flowing from the extracranial artery into the intracranial artery are anastomosed at the base of the brain via the anterior and posterior communicating arteries. When one of these vessels is occluded, the other circle of Willis arteries function as collateral blood vessels to supply cerebral blood flow to the volume normally supplied by the stenotic occluded vessel. The third route involves anastomosis of the cerebral pia mater artery. When the middle cerebral artery, anterior cerebral artery, or posterior cerebral artery are occluded distal to the circle of Willis arteries, the region normally supplied by these arteries is perfused via anastomosis of the cerebral pia mater arteries within the surrounding perfusion area. The development of collateral circulation varies greatly from person to person, but it is thought that the rate of progression of cerebrovascular disease, such as stenosis occlusion, influences the development. The details of cerebrovascular diseases will be discussed in detail in Chap. 7.

3.3.5 Gray Matter and White Matter Have Different Capillary Densities

The brain is composed of gray matter, which mainly consists of the cell bodies of neurons, and white matter, which consists of the nerve fibers (axons) that connect the neurons. The gray matter is called gray matter because the neurons appear dark and gray. The white matter is called white matter because the myelin sheath that surrounds the nerve fibers appears white. In humans, gray matter is approximately a few millimeters thick and covers the surface of the brain, which is called the cerebral cortex. The gray matter and white matter take up approximately the same volume of the cerebrum. In the center of the cerebrum are the thalamus, interneurons, basal ganglia, and brainstem, which are important nerve nuclei closely involved in vital functions. The basal ganglia also contain the paths that connect the thalamus, which relays peripheral nerves in the trunk region, to the cerebral cortex. The nerve nuclei are made up of neuron of gray matter and are sometimes referred to collectively as deep gray matter.

The density of capillaries differs between the gray matter consisting of neurons and the white matter consisting of nerve fibers. The energy consumption of the gray matter is much greater than that of the white matter, and hence the cerebral blood flow supplied to the gray matter is also greater. In animal experiments, cerebral blood flow measured with autoradiography (see Sect. 6.4.2) is 100 mL/100 g/min in gray matter and 20 mL/100 g/min in white matter (Fig. 3.12) (Reivich et al. 1969). It has been confirmed by simulation and MRI measurements that the density of capillaries is in proportion to the distribution of cerebral blood flow (Ostergaard et al. 1999). Measurements of the cumulative length of the capillaries in a 1 mm^3 volume of the cat brain found it to be approximately 800 mm/mm^3 for gray matter and approximately 350 mm/mm^3 for white matter (Purves 1972). In various experiments, the ratio of cerebral blood flow in gray matter to white matter has been measured to be 4:1. In humans, the average cerebral blood flow in the cerebrum is estimated to be 50 mL/100 g/min, which is obtained as the average of 80 mL/100 g/min for gray matter and 20 mL/100 g/min for white matter, assuming that gray matter and white matter have the same volume (Lassen 1985).

Capillaries are distributed throughout the tissue of the entire brain. In the gray matter, the distance between capillaries is approximately 50 μm, which is the distance that oxygen can diffuse from the blood vessels while the oxygen partial pressure necessary for brain tissue is maintained. The oxygen partial pressure is based on "Krogh's cylinder" (Fig. 3.13), where the oxygen partial pressure decreases along the longitudinal direction of blood flow from the arterial end to the venous end of microvessels, and outwards from the vessel

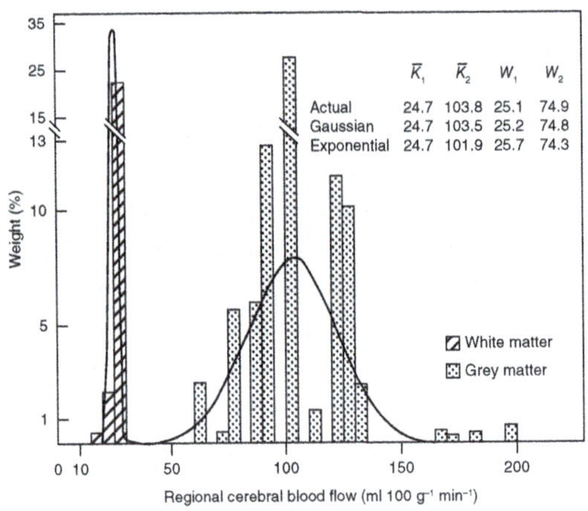

Fig. 3.12 Regional cerebral blood flow (CBF) distribution in the white and gray matter in cat: CBF was measured by autoradiography with ^{14}C-labeled iodoantipyrine. The white matter and gray matter were 20 mL/100 g/min and 100 mL/100 g/min, respectively. (Reivich M, In Cerebral Blood Flow 1969, ed Brock M et al, Springer Berlin)

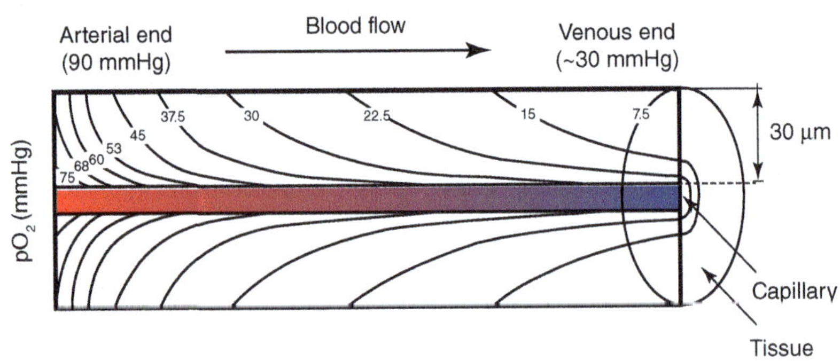

Fig. 3.13 Krogh's cylinder: The isobaric distribution of partial pressure of oxygen along the longitudinal and radial directions of the capillaries assuming uniform oxygen consumption in the brain tissues around the vessel

center along the radial direction (Krogh 1919). The axial and radial oxygen partial pressure gradients of Krogh's cylinder are determined according to the metabolic rate of oxygen consumption in the tissue surrounding the vessel. The spacing of the capillaries is determined accordingly. Therefore, in areas with low cerebral oxygen metabolism, such as the white matter, the oxygen partial pressure gradient is less steep, so the radius of Krogh's cylinder effectively becomes larger and the capillary spacing widens.

3.4 Oxygen Delivery by Red Blood Cells

Red blood cells contain hemoglobin and have an important role in transporting oxygen. They move through the capillaries, of which the inner diameter is approximately 5 μm and narrower than the diameter of red blood cells. Hence, red blood cells rub against the capillary wall. A healthy red blood cell is easily deformed and can move freely through the capillaries. When red blood cells are deformed, their lipid-rich cell membranes increase the area of direct contact with the endothelial cells of the vessel wall, making it easier for lipid-soluble O_2 and CO_2 gases bound to the hemoglobin of the red blood cells to move through the vessel wall.

3.4.1 Oxygenation and Deoxygenation of Hemoglobin

Hemoglobin is a red pigment that contains four iron atoms per molecule, and it binds and dissociates with four molecules of O_2 to transport O_2 from the lungs to the body tissues. At the same time, hemoglobin removes CO_2 from body tissues to the lungs. The physical properties of hemoglobin differ greatly between oxygenated hemoglobin, which binds O_2, and deoxygenated hemoglobin, which dissociates O_2. First, they differ in their absorbance properties. Deoxygenated hemoglobin is dark red, while oxygenated hemoglobin is bright red. This difference in absorbance properties brings unexpected benefits to the measurement of brain function. Using this difference in absorbance characteristics, a near-infrared spectrometer has been developed to monitor brain function based on changes in cerebral blood volume (see Sect. 6.3.4). In addition, the magnetic properties of hemoglobin are different depending on whether it is oxygenated or deoxygenated. Deoxygenated hemoglobin is paramagnetic, while oxygenated hemoglobin is diamagnetic (Fig. 3.14). This difference in magnetic properties, in conjunction with MRI, has been of great benefit to brain science (see Sect. 8.2.1). Functional brain mapping with MRI is a completely noninvasive technique that has revolutionized brain science.

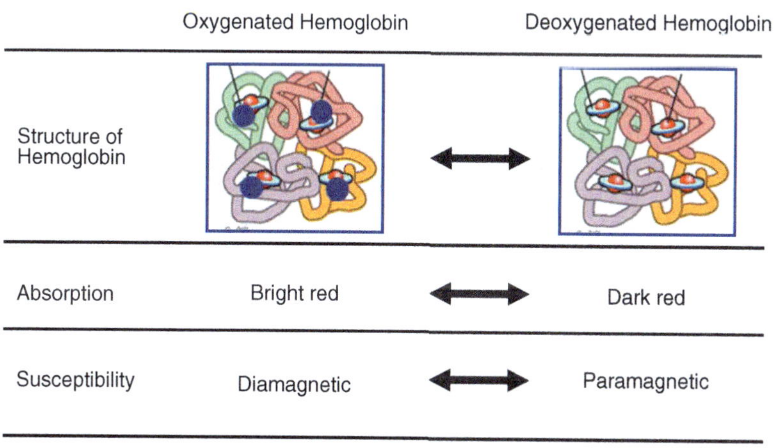

Fig. 3.14 Comparison of the physical properties of oxygenated and deoxygenated hemoglobin: Oxygenated hemoglobin contains four molecules of oxygen (blue) (left side) and deoxygenated hemoglobin has no oxygen (right side). The two forms have different physical properties in terms of light absorption and magnetic susceptibility

3.4.2 Role of the Pulmonary Circulation and Systemic Circulation

One of the major functions of blood flow is to take in O_2 from the lungs and supply it to body tissues. For this purpose, the heart pumps blood into the pulmonary system before it is pumped into systemic circulation (Fig. 3.15). In pulmonary circulation, blood is sent from the right ventricle through the pulmonary artery to the lung and returns through the pulmonary vein to the left atrium. O_2 is taken up in the alveoli of the lung. The pulmonary circulation is also called *minor circulation* because the blood flow path is smaller than that of systemic circulation. On the other hand, in systemic circulation blood is sent from the left ventricle through the arteries to the whole body, including the brain and the muscles of the trunk, and returns as the venous blood to the right atrium. O_2 is delivered via systemic circulation to the body and released from the blood in the capillary region. The systemic circulation is called the *major circulation* because the blood circulates through a large flow path. The blood that has taken in O_2 in the lungs is called arterial blood containing O_2, while the blood that has released O_2 in the body tissues is called venous blood.

In contrast to O_2, CO_2 is collected in the systemic circulation and released in the pulmonary circulation. In other words, CO_2 is collected from tissues throughout the body, returned to the right atrium as venous blood, and then pumped from the right ventricle to the lungs in the pulmonary circulation, where it is released in the alveoli. In addition, waste products are also collected in the systemic circulation and filtered out through the kidneys before the blood returns to the right atrium.

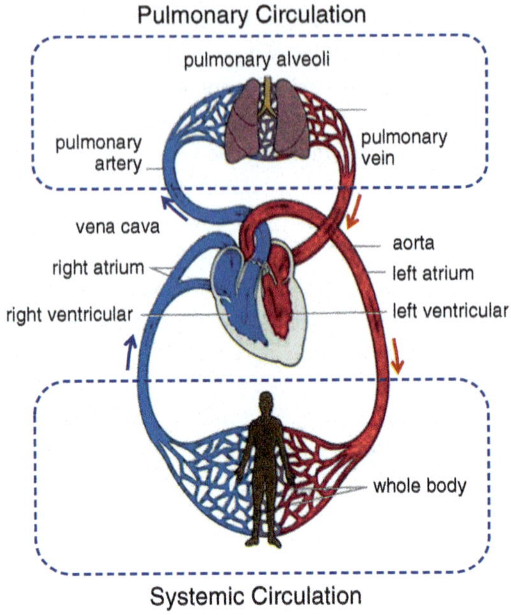

Fig. 3.15 Pulmonary circulation and systemic circulation: Blood circulation consists of the pulmonary circulation (minor circulation), which carries out gases exchange in the lungs, and the systemic circulation (major circulation), which delivers oxygen and nutrients to the whole body

3.4.3 Oxygen Dissociation Curve

The binding and dissociation of O_2 to hemoglobin depends on the partial pressure of oxygen around red blood cells. The relationship between oxygen partial pressure and oxygen saturation is called the oxygen dissociation curve, which has a nonlinear S-like (sigmoidal) shape (Fig. 3.16). The oxygen dissociation curve shifts left or right depending on the CO_2 concentration, pH, temperature, and other factors in the vicinity of hemoglobin. The oxygen saturation on the vertical axis is the ratio of oxygenated hemoglobin to total hemoglobin. The oxygen content of the blood is the sum of the oxygen bound to hemoglobin and the dissolved oxygen in the plasma.

The partial pressure of oxygen in the alveoli is approximately 140 mmHg, which is 21% of the atmospheric pressure of 760 mmHg. However, it reduces to approximately 100 mmHg after the exchange that supplies O_2 to the red blood cells of arterial blood in the alveoli. Hemoglobin binds and takes up oxygen in an environment (e.g., in the alveoli) where the O_2 partial pressure is high (rich O_2). The total area of the alveoli in contact with the blood is 60 ~ 80 m^2 (equivalent to half the surface of a tennis court), which is a vast area

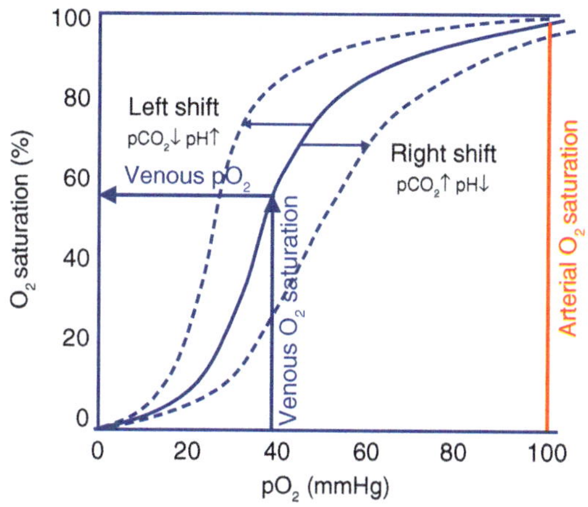

Fig. 3.16 Oxygen dissociation curve: Relationship between the oxygen saturation of hemoglobin (Y-axis) and the partial pressure of oxygen (X-axis). In body tissues at low oxygen partial pressure hemoglobin dissociates and releases oxygen, while in alveolar regions at high oxygen partial pressure hemoglobin absorbs and binds oxygen. The oxygen dissociation curve shifts from side to side depending on the pH and CO_2 partial pressure of the blood

promoting gas exchange between the blood and the atmosphere. Furthermore, the thickness of the alveolar wall is very thin (0.2 ~ 0.6 μm), so diffusion exchange of the gas in the alveoli with the gas in the capillaries can be efficiently performed. On the other hand, in an environment (e.g., body tissues) with low O_2 partial pressure (poor O_2), O_2 is dissociated (oxygen is released) from hemoglobin. The oxygen dissociation curve reflects the characteristics of O_2 attachment to and detachment from hemoglobin according to the surrounding O_2 partial pressure.

The arteries containing arterial blood that has taken up O_2 in the alveoli eventually reach the body tissues after repeated branching, and because the O_2 partial pressure is low in the body tissues, the O_2 in the blood diffuses rapidly into the surrounding area and gradually reaches equilibrium with the O_2 partial pressure in the most peripheral areas. At this point, the O_2 partial pressure is approximately 40 mmHg so the oxygen saturation drops to around 70%. The O_2 partial pressure drops to approximately 20–40 mmHg in the blood that has finished supplying O_2 to the tissues in the peripheral part of the systemic circulation, which means that the oxygen saturation drops to approximately 50% (Table 3.1). The slope of the oxygen dissociation curve becomes steeper at this point, indicating that hemoglobin dissociates bound O_2 more rapidly to facilitate the release of oxygen to the tissues in response to the decrease in O_2 partial pressure corresponding to the demand for O_2 in the peripheral area. In other words, an increase in oxygen metabolism in the peripheral area decreases the O_2 partial pressure accordingly, and oxygenated hemoglobin dissociates O_2 in response to supply body tissues. Because the slope of the curve is steep at low partial pressures, a slight decrease in the partial pressure of O_2 promotes oxygen dissociation from oxygenated hemoglobin, rapidly lowering the oxygen saturation and facilitating the supply of more O_2 to surrounding tissue.

3.4.4 Hematocrit

Blood is made up of liquid components like plasma and solid components like red blood cells. Most of the solid components are red blood cells, and the ratio of red blood cells to the total blood volume is called the hematocrit. It reflects the concentration of hemoglobin in the total blood. In general, the hematocrit ranges from 0.3 to 0.5, but it differs between men and women, with women having a slightly lower value than men.

As mentioned earlier, red blood cells are heavier than plasma, so they tend to gather in the center of blood vessels, where the flow velocity is faster, and the plasma components flow closer to the periphery. This is called the *sigma effect*. As a result of the *sigma effect*, when a small artery, such as a perforating artery, branches off at right angles from a large cerebral surface artery, the composition of blood flowing in the branch artery is relatively higher in plasma than that of blood in the surface artery. This is called the Fahraeus-Lindqvist effect. The hematocrit, which indicates the percentage of red blood cells, decreases as one moves toward the peripheral blood vessels due to the Fahraeus-Lindqvist effect. The hematocrit in the microvasculature of brain tissue is 60–80% of the hematocrit in the large vessels of the trunk. The apparent concentration of red blood cells decreases as they move into the peripheral blood vessels. Hematocrit is also an important parameter for microcirculation hemodynamics and vascular resistance.

References

Attwell D, Mishra A, et al. What is a pericyte? J Cereb Blood Flow Metab. 2016;36(2):451–5. Review.

Dalkara T, Gursoy-Ozdemir Y, Yemisci M. Brain microvascular pericytes in health and disease. Acta Neuropathol. 2011;122(1):1–9. Review.

Dienel GA. Astrocytic energetics during excitatory neurotransmission: what are contributions of glutamate oxidation and glycolysis? Neurochem Int. 2013;63(4):244–58. Review.

Hartmann DA, Underly RG, et al. Pericyte structure and distribution in the cerebral cortex revealed by high-resolution imaging of transgenic mice. Neurophotonics. 2015;2(4):041402.

Iadecola C. The neurovascular unit coming of age: a journey through neurovascular coupling in health and disease. Neuron. 2017;96(1):17–42.

Klatzo I, Piraux A, Laskowski EJ. The relationship between edema, blood–brain-barrier and tissue elements in a local brain injury. J Neuropathol Exp Neurol. 1958;17(4):548–64.

Krogh A. The number and distribution of capillaries in muscles with calculations of the oxygen pressure head necessary for supplying the tissue. J Physiol. 1919;52(6):409–15.

Lassen NA. Normal average value of cerebral blood flow in younger adults is 50 ml/100 g/min. J Cereb Blood Flow Metab. 1985;5(3):347–9.

Masamoto K, Takuwa H, et al. Microvascular sprouting, extension, and creation of new capillary connections with adaptation of the neighboring astrocytes in adult mouse cortex under chronic hypoxia. J Cereb Blood Flow Metab. 2014;34(2):325–31.

Ostergaard L, Chesler DA, et al. Modeling cerebral blood flow and flow heterogeneity from magnetic resonance residue data. J Cereb Blood Flow Metab. 1999;19(6):690–9.

Purves MJ. Chap. 2, pp. 26–27. In The physiology of the cerebral circulation. Monograph Physiol Soc. 1972;(28):1–414.

Reivich M, Slater R, Sano N. Further studies on experimental models of cerebral clearance curves. In: Brock M, Fieschi C, Ingvar DH, Lassen NA, Schürmann K, editors. Cerebral blood flow. Berlin: Springer; 1969. p. 4–7.

4 Measuring Cerebral Blood Flow: History

Abstract

This chapter considers the history of cerebral blood flow measurement. The first study of cerebral hemodynamics was by Mosso in the late nineteenth century. Using a special self-constructed device on subjects who had suffered head trauma, he found that brain volume changed when mental activity occurred. A half century later, Kety developed a method to measure cerebral blood flow using an inert gas as a diffusible tracer that accumulated in the brain after inhalation. Lassen extended Kety's method to use radioisotopes. The development of tomography made noninvasive quantitative measurements of brain function possible. Accurate measurements of cerebral blood flow became very important for the clinical diagnosis of various neural diseases as well as for understanding fundamental neurophysiology. PET with radioactive tracers such as oxygen-15 gas and fluorine-18 FDG contributed greatly to knowledge of brain energy physiology and circulatory metabolism. The many technological and scientific breakthroughs made by our predecessors are briefly described here.

Keywords

Brain volume · N_2O inhalation · Xe-133 clearance · Inert gas · Radioactive tracer · Mean transit time · High-speed SPECT

4.1 The Early Days of Cerebral Blood Flow Measurement

It was known by the eighteenth century that mental activity is located in the brain. While phrenology was discredited by the late nineteenth century, it was in a way the forerunner for current theories where any brain functions are localized to specific regions.

4.1.1 Measuring Changes in Brain Volume

The first person to report that brain volume changes with brain activity was an Italian physiologist named Mosso (Fig. 4.1). He found that the brain volume of patients with skull defects caused by trauma changed from time to time. Mosso made a device to record these changes and with it he measured the change in brain volume when the patient was given a task, such as a calculation, that required the brain to work (Fig. 4.2) (Mosso 1880). Later, Roy & Sherrington (Fig. 4.3) reported that brain volume changed in response to sciatic nerve stimulation when measured with a brain volumeter, or oncometer, they had developed for animals (Fig. 4.4) (Roy and Sherrington 1890). Both Mosso and Roy & Sherrington actually measured the change in cerebral blood volume with brain activity. These were the first measurements showing the link

I. Kanno, *What Cerebral Blood Flow Tells Us*, https://doi.org/10.1007/978-981-97-6382-5_4

between brain activity and cerebral blood volume, which is the current basis of functional brain imaging. After drilling a hole in the cranium, Roy & Sherrington also showed that stimulation of the peripheral nerves increases blood flow in the brain and results in an increase in intracranial volume. This was the first study to observe cerebral blood flow change after nerve stimulation.

Fig. 4.1 Angelo Mosso (1846–1910)

4.1.2 Early Attempt to Quantify Cerebral Blood Flow

Sometime after the work of Roy & Sherrington, attempts were made to quantify blood flow. The first attempts to measure cardiac output were performed by Stewart (1921) and Hamilton (1931) in the early 1900s (also see the review by Schlant et al. 1959). Using a rabbit, ink was injected into the vein and the earlobe, through which blood vessels could be seen, was observed as the ink traversed the vessels. Light that shone through the ear and onto a detector on the other side was used to measure the concentration of ink in the blood. In this way, the mean transit

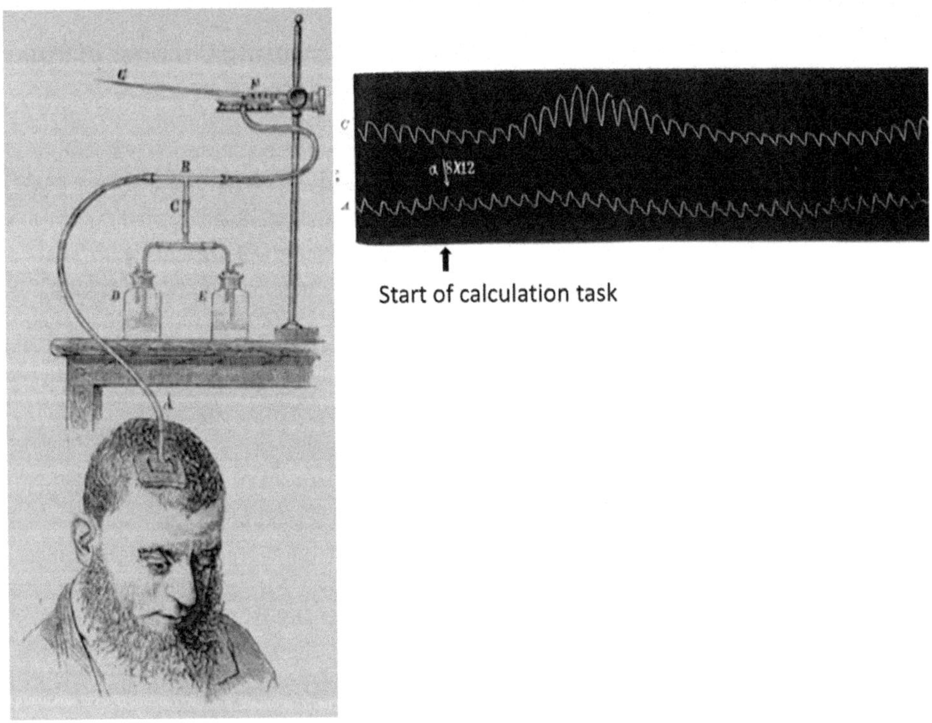

Fig. 4.2 Mosso's cranial cavity recording device: Mosso found that the brain volume changed for patients with a hole in the skull caused by injury. He recorded the swelling of the brain when the patient was asked to do mental arithmetic. (Mosso A, Accademia dei Linsei 1880)

time (MTT) of the ink was measured and used to estimate the cardiac output. This same technique was also used in later studies to estimate the blood flow of individual organs and tissue. However, to quantitatively calculate the blood flow it is necessary to know the volume of blood vessels in the organ. There are also many other unknowns that prevent quantitative measurement per organ. Nevertheless, the Stewart-Hamilton method was an important step towards quantifying blood flow.

Fig. 4.3 Roy CS (1854–1897) (right) and Sherrington CS (1857–1952) (left)

4.2 Quantification of Cerebral Blood Flow Using Inert Gases

Approximately 60 years after Mosso, a mathematical model based on Fick's principle (Fick 1881) was developed by Kety & Schmidt for estimating cerebral blood flow. Using the model and laughing gas (N_2O) as a tracer, they were able to accurately measure cerebral blood flow for the first time.

4.2.1 Quantification with Laughing Gas

Kety (Fig. 4.5) and Schmidt quantitatively determined cerebral blood flow using N_2O, which is an anesthetic gas that is commonly referred to as laughing gas (Kety and Schmidt 1945). N_2O can freely diffuse out of the blood vessels into brain tissue but because it is chemically inert, it is not absorbed into brain tissue, returns to blood vessels, and eventually washes out again. The mean transit time of N_2O in the circulation was accurately determined from the difference of concentrations in arterial and venous blood, and cerebral blood flow was estimated from the rate of tracer saturation into the whole-brain based on Fick's principle (Fick 1881). Kety & Schmidt were the first to suc-

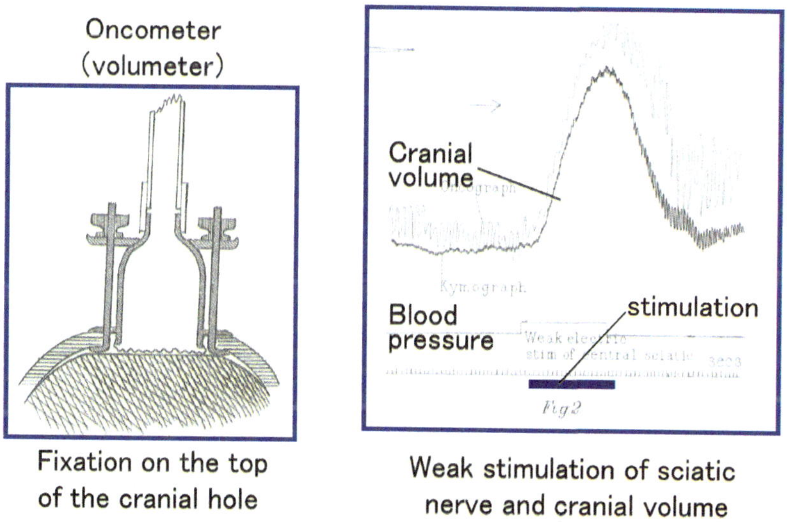

Fig. 4.4 Roy and Sherrington's cranial volumetric oncometer and measurements: The upper cranium of a dog was removed and the oncometer was fixed to the top of the head. Observations in this condition showed that the intracranial volume increased when the sciatic nerve was stimulated. (Roy CS & Sherrington CS J Physiol 1890)

cessfully quantify whole-brain cerebral blood flow (Kety and Schmidt 1945, 1948). Using an inert gas as a diffusible tracer and calculating cerebral blood flow from the rate of tracer saturation were important steps forward in developing the technology to measure cerebral blood flow.

Fig. 4.5 Seymour S Kety (1915–2000)

The N_2O method requires a puncture behind the ear to draw venous blood from the internal jugular vein at the base of the skull, and a puncture into the brachial artery to withdraw arterial blood. Both these procedures are extremely invasive and painful and would not be ethically acceptable today (Fig. 4.6). Nevertheless, this was the first method that could quantitatively measure whole-brain cerebral blood flow, had high measurement reproducibility and accuracy, and could be used to measure cerebral blood flow during sleep and other physiological conditions. It was a groundbreaking method that raised the profile of cerebral blood flow to a scientific level for the first time. Moreover, this method is excellent in that it measures the difference in oxygen content of venous and arterial blood samples, and multiplying this difference by the whole-brain cerebral blood flow can be used to estimate the cerebral metabolic rate of oxygen (CMRO2) of the brain. Thus, Kety & Schmidt were the first to measure cerebral oxygen metabolism for the whole brain. Even in his later years, Kety attended the meeting of the International Conference on Cerebral Blood Flow and Metabolism. Although

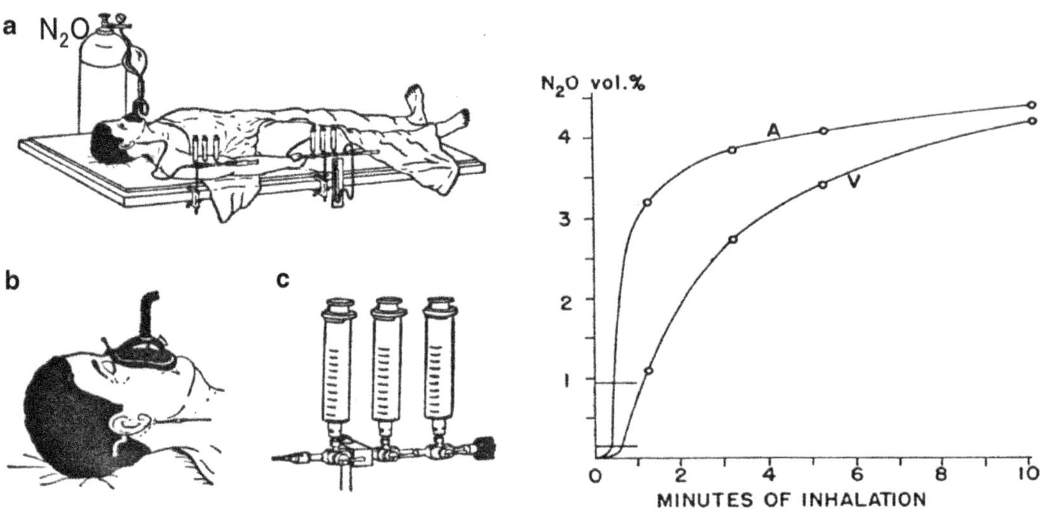

Fig. 4.6 Kety & Schmidt's N_2O saturation method: N_2O is inhaled via a closed mask. The arterial blood and venous blood were sampled from the peripheral artery and jugular veins, respectively. The concentration of N_2O in both arterial blood and venous blood was measured. From the arteriovenous differences, the cerebral saturation curve of N_2O was measured and the cerebral blood flow was calculated based on Fick's principle. (Kety SS, Schmidt CF, J Clin Invest 1948)

he seldom spoke, he was always seated in the front and listened quietly. Lassen, one of Kety's students, continued to invite Kety to his research conferences even in the 1990s.

4.2.2 Anesthetic Effects of N_2O

As well as being extremely invasive, Kety-Schmidt's N_2O saturation method had another major problem: the N_2O had an anesthetic effect. This meant that the subject was feeling anesthetized and not in a strictly awake, resting state when the measurement was performed. To measure with high sensitivity, it is desirable to have the subject inhale a large amount of N_2O, but if this is done, the anesthetic effect on the subject is stronger and the reproducibility of the measurement would be affected. Although the goal was to measure cerebral blood flow in the awake state, it was likely that the cerebral blood flow in the anesthetized state was measured in some cases. In addition, if the subject suffered from some cerebrovascular disorder or other disease, the anesthesia was significantly invasive and required time for the patient to recover. Furthermore, Kety-Schmidt's method caused the accumulation of N_2O in the body tissues, and it took a long time for the tracer to be discharged from the body, making it difficult to repeat measurements.

4.2.3 Breakthrough by Introducing Radioactive Tracers

These limitations of Kety-Schmidt's method were solved successfully by Lassen & Ingvar in the 1960s. Instead of N_2O, they used a radioisotope (RI) of an inert gas as a tracer for cerebral blood flow measurement (Lassen and Ingvar 1961). Either krypton-85 or xenon-133 was dissolved into saline and was administered by quickly injecting into the internal carotid artery. This led to several breakthroughs in cerebral blood flow measurement: (1) The use of RIs made it possible to measure cerebral blood flow regionally from outside the skull; (2) Rather than observing the buildup of tracer in the tissue, as in Kety-Schmidt's technique (integral method), the measurement was changed to observing the clearance of the RI from tissue (differential method) (Fig. 4.7) (Lassen and Hoedt-Rasmussen 1966); (3) Since the tracer is administered within a few seconds, it could be assumed that the input function (see Sect. 5.4.4) was a unit-impulse and cerebral blood flow could be calculated directly from the exponential clearance curve (Zierler 1965); (4) Only a small amount of inert gas is needed as a tracer so that the anesthetic effect, which cannot be avoided with N_2O, was eliminated.

Knowing that cerebral blood flow fluctuates physiologically, Lassen developed a clearance method that could be repeated in a short time compared to Kety's saturation method. Thus, compared to the N_2O saturation method, Lassen's rapid xenon-133 RI tracer clearance method killed four birds with a single stone: i.e. shorter measurement time, repeatable measurements, substantially reduced anesthesia effect, and no need for arterial blood sampling. However, it was still invasive, requiring an arterial puncture for injection of the tracer into the internal carotid artery. Improvements were subsequently made with the development of a noninvasive xenon-133 inhalation method, which allowed measurement of cerebral blood flow in both hemispheres. For almost 40 years up until about the year 2000, the xenon-133 inhalation method was the only way to quantitatively measure cerebral blood flow at non-PET facilities. As the only quantitative measure of cerebral blood flow, it was an indispensable tool for neurosurgeons in the evaluation of patients for bypass surgery and other neurosurgical treatments for cerebrovascular diseases.

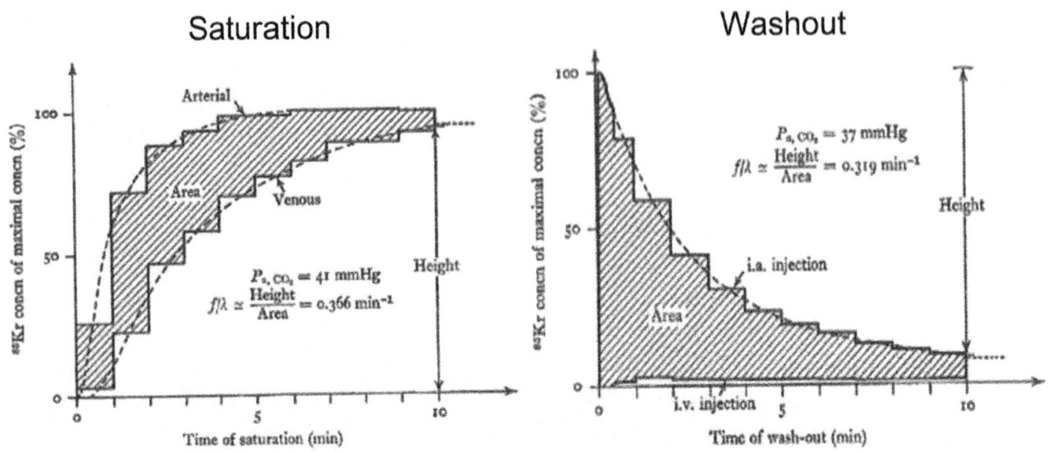

Fig. 4.7 The N_2O saturation method and the krypton-85 RI washout method: The change from the integral method (left) to the differential method (right) shortened measurement time and allowed repeated cerebral blood flow measurements. (Lassen NA, Hoedt-Rasmussen K, Circ Res 1996)

Column: An Anecdote About the Xenon-133 Brain Clearance Method

David Ingvar, who developed the xenon-133 brain clearance method with Niels Lassen, visited the Akita Brain Institute around 1978. When he learned that the author had studied with Lassen a few years earlier, he told me an anecdote about the time when he and Lassen were working on the xenon-133 brain clearance method. Lassen was at Bispebjerg Hospital in Copenhagen, Denmark, and Ingvar was at Lund University in southern Sweden, 1 ½ hours away by ferry and train. They had both been obsessed with the development and application of regional cerebral blood flow measurement since their days at the National Institutes of Health (NIH) in the United States in the 1950s. They first used krypton-85, a RI with fewer gamma rays than xenon-133, but finally managed to obtain and test xenon-133. Since the gamma energy of xenon-133 is different from that of krypton-85, the detector amplifier had to be adjusted for xenon-133 measurements. The adjustment took a long time, and it was after 5 pm when the detector was finally ready. The scheduled patients and other hospital staff had already gone home. The detector was ready and the xenon-133 was available, but Lassen and Ingvar were the only ones left. They did not want to waste any precious xenon-133 so they tossed a coin and the winner (unfortunately, the author doesn't remember which) injected xenon-133 saline into the other's internal carotid artery and took a measurement. This was the story of the first measurement with the xenon-133 brain clearance method.

4.2.4 The Development of the Xenon-133 Method

Niels Lassen's sense of mathematics was astonishing. Any electrical engineer, including the author, would have used Laplace's transform to solve the differential equation for tracer behavior, but Lassen solved the problem intuitively. Lassen's students said that he had a "diamond head", which is a phrase that has two meanings. They meant that he had a brilliant brain that shone like a diamond, but it was also true that he was very stubborn, so his head was as hard as diamond. Once he believed in something, he would never change his mind no matter what others said. However, he was very curious and whenever he heard of a new discovery somewhere in the world, he would immediately fly to that place to check it out for himself. It is still surprising to the author that he did not invent PET in his lifetime. It might have been because of his insistence on using xenon-133 and its single-photon emission computed tomography (SPECT) method to measure cerebral blood flow, which he had established himself, or because he did not want to deal with the huge amount of personnel, equipment and complexity required for a PET facility, but the real reason remains a mystery.

Column: The Humanity of Niels Lassen

It was early spring in Copenhagen when I was invited to a study group organized by Lassen. As we strolled through Dyrehaven Park in the suburbs, he began to tell me that the surrounding grasses could only enjoy the sunlight at that time of the year. He told me that in a little while the leaves of the trees above us would grow thicker and the rays of sunlight would be blocked by the taller trees and would not reach the ground. I was surprised to hear such a story for the first time from a man with whom I had only previously had academic conversations. When I asked Lassen's students, they were also surprised that Niels talked about such things. Early in my studies in Denmark, he invited my family to his home. He had just gotten divorced and treated us to a simple meal of roasted chicken. After dinner, we were busy playing Lego with the kids, and he talked to me while pointing at his head saying that this (Lego) is all we have in Denmark. I think he was trying to say that ideas are the only thing the Danes have. At that time (1978), he was in his fifties and had a busy schedule traveling around the world. Nevertheless, he invited us and spent time with us. He was the head of the department at his hospital, and the nurses and technicians at the hospital, who did not know of his academic greatness, were always wondering why so many guests were coming to visit Niels.

4.2.5 Introduction of the Xenon-133 Method into Japan

Let us return in time to 1970. After Lassen et al. reported their method for measuring regional cerebral blood flow with xenon-133 gas, my boss Kazuo Uemura (Fig. 4.8) at the Akita Cerebrovascular Research Center (Akita Brain Institute) was the first to introduce the

Fig. 4.8 Kazuo Uemura (1935–2004)

method to Japan. Uemura made a six-channel xenon-133 detector that divides the temporal lobe into six ≈ 5 cm diameter sections in a 2 × 3 grid, enabling the xenon-133 brain clearance method of Lassen et al. The xenon-133 saline (xenon-133 gas was dissolved in a solution of 1–2 mL of saline) was injected in about 1–2 seconds into the internal carotid artery, and the xenon-133 brain clearance curve was measured with the six-channel detector placed on the temporal lobe (Fig. 4.9). The position of the detector for each xenon-133 brain clearance curve was determined, and the local cerebral blood flow was calculated from the data for each detector. Following the paper by Lassen et al., the height-over-area (H/A) method (Zierler 1965) was used to estimate cerebral blood flow from the full 10-minute brain clearance curve, and the initial slope (IS) method was used to estimate the blood flow from the logarithmic slope of the initial 2 minutes of the data (Fig. 4.10). Uemura later

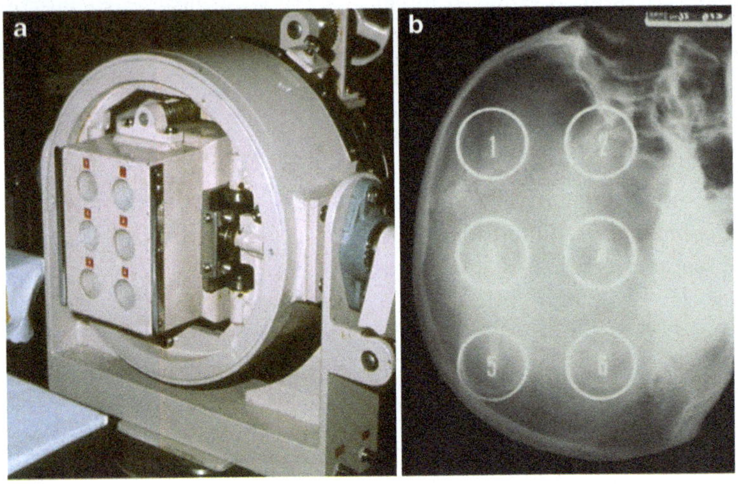

Fig. 4.9 The six-channel detector for the xenon-133 clearance method: (**a**) The six-channel detector, made from lead, covered the hemisphere of the head with six detectors spaced 5 cm apart. (**b**) X-ray image of the six-channel detector superimposed on a lateral view of the head. This confirmed the detector location on the subject's head

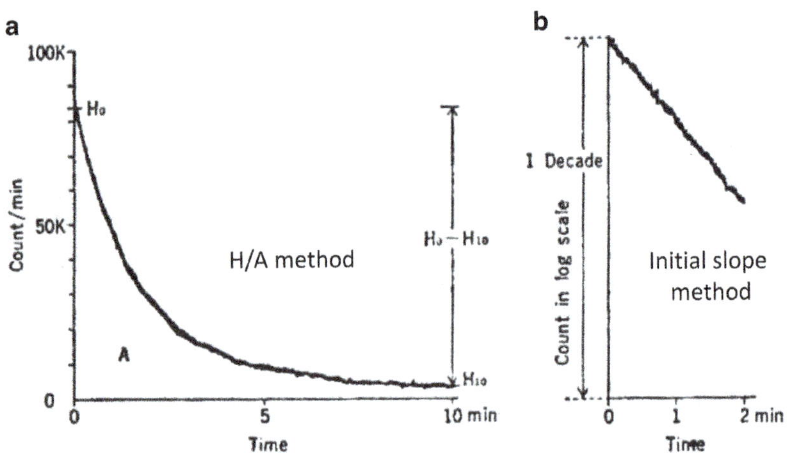

Fig. 4.10 Graphical method for calculating cerebral blood flow from the xenon-133 clearance curve: (Left) The Height over Area (H/A) method calculates cerebral blood flow by dividing the initial height (H) of the clearance curve by the 10-minute area (A). (Right) The Initial Slope (IS) method calculates cerebral blood flow from the logarithmic slope of the initial two minutes of the clearance curve

improved his measurements of cerebral blood flow by introducing a 16-channel detector to measure in finer subregions (approximately 2 cm diameter), and eventually he was able to perform cerebral blood flow measurements at a resolution of around 1 cm with a gamma camera (Autofluoroscope). We vigorously promoted the development of measurement devices (Fig. 4.11). However, we later heard that neurosurgeons and neurologists on the clinical side complained that it was difficult to follow up the cerebral blood flow of patients before and after because of the rapid evolution of our measurement devices.

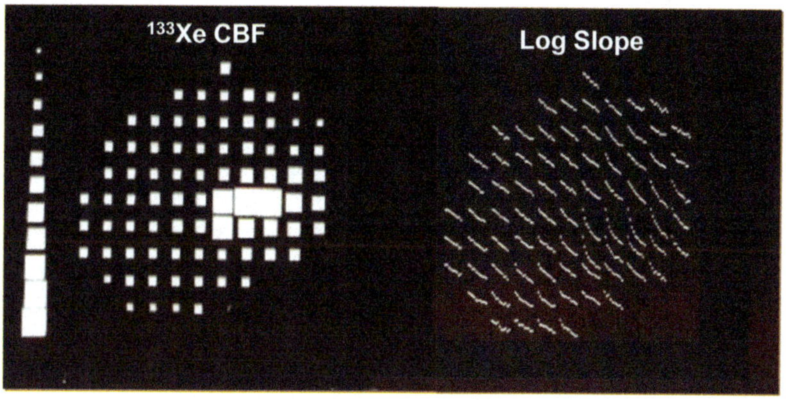

Fig. 4.11 Cerebral blood flow measured with an Autofluoroscope: A patient with middle cerebral artery infarction was examined with the xenon-133 clearance method. (Left) Hemispheric map of cerebral blood flow calculated with the H/A method. (Right) Logarithmic slopes during the initial two minutes. Both maps were recorded on Polaroid film using a camera with its shutter open while pointed at the moving spots on a CRT display. The Autofluoroscope had a 14 by 21 array of NaI-crystal detectors with an interval of 1 cm in both directions

Column: Digital Computers in the 1970s

The computer used to analyze the xenon-133 brain clearance curve was a "digital computer," as opposed to the mainstream analog computers of the time. It cost more than a house (Fig. 4.12). The memory consisted of a few millimeter-diameter ring-shaped core magnets supported by orthogonal wires for each bit, and 20 of them were lined up in a row to form a word. The core contained a network of orthogonal wires, with the vertical wires used to read 0 and/or 1, and the horizontal wires used to write 0 and/or 1. The external storage device was a magnetic drum (a vertical

Fig. 4.12 A computer in the 1970s (JEOL-JEC7): It contained an 8 kw (kiloword) internal core memory and 16 kw external magnetic drum memory. Data and programs were loaded from a magnetic tape. The programs to analyze xenon-133 clearance curve data and to calculate cerebral blood flow were coded in machine and assembler languages

40–50 cm diameter cylinder that rotated) with a storage capacity of 16 kilowords. As there was no BASIC, FORTRAN, or C, the only programming languages were machine language and assembler language. Using a combination of these, we wrote a program that read in a paper tape containing the radioactivity of the xenon-133 brain clearance curve (AD-converted into binary numbers from the data recorded on a magnetic tape), logarithmically converted it, calculated the cerebral blood flow from the gradient using the least-squares method, and printed it out on an IBM Dharma-type printer.

4.2.6 Limitations of Xenon-133

The xenon-133 brain clearance method has a poor image resolution of 1 cm at the brain surface. This is due to the low energy (81 keV) of xenon-133 gamma rays, which have only a short penetration distance because they are easily absorbed and attenuated by biological tissue, resulting in many scattered rays. In addition, as tomography requires high-speed dynamic measurement the signal-to-noise ratio (S/N ratio) of the SPECT data was low, so we were facing a major problem using xenon-133.

For clinical measurements of cerebral blood flow that do not require absolute quantification, the xenon-133 brain clearance method has been replaced by a method that uses a technetium-99m-labeled chemical microsphere tracer that became commercially available in 2000. Since the technetium-99m-labeled chemical microspheres easily pass through the BBB and remain in place around the capillaries in proportion to the cerebral blood flow (see Sect. 5.2.4), measurements can be made with a standard SPECT system. Thus, a high-speed SPECT system is unnecessary. Since sufficient counts can be obtained in a relatively short time, it is now possible to measure high signal-to-noise images with just a gamma-camera SPECT system.

4.3 Tomography

Even though the brain is covered by the skull it has long been known that many complex functions are localized to the brain. There has therefore been a longstanding desire to see inside and examine the function of the living brain.

4.3.1 The Dawn of Tomography

Attempts to examine the inside of the brain using RIs have been performed since the 1950s. Brownell and colleagues at Massachusetts General Hospital (MGH) measured tomographic images using two opposing plane detectors and coincidence counting. In the 1960s, Yamamoto and colleagues at Brookhaven National Laboratory (BNL) developed a helmet-arrayed positron detector to measure cerebral blood flow and attempted tomography.

David Kuhl at the University of Pennsylvania (Fig. 4.13) developed a gamma-ray detector that rotates around the head while scanning. The Mark IV system his group produced in 1976 could measure a full-scale SPECT image of the brain (Kuhl et al. 1976). However, the Mark IV took more than 10 minutes to collect a single data set. Since Kuhl and his group focused on measuring brain pathology, they were less interested in resolving signal dynamics at the expense of signal-to-noise, so their system did not use the xenon-133 clearance method to improve measurements of cerebral blood flow.

Fig. 4.13 David E. Kuhl (1929–2017)

Column: David Kuhl Maintained the Device He Constructed

David Kuhl was an old-fashioned doctor of nuclear medicine who developed his own device for his measurements. I had a chance to talk with David Kuhl several times at conferences and other gatherings. When I talked to him about the difficulties I had in developing a tomography system that rotates the patient chair in front of an Autofluoroscope, David Kuhl immediately understood my frustration. I imagined this was probably because at that time he was struggling to repair troubles with his own system, the Mark IV quadrangle SPECT, with a screwdriver. After that, we had several opportunities to meet each other at various international conferences (sometimes he brought his wife with him), and we enjoyed talking about many things. David Kuhl won the Japan Prize in 2009.

When Lassen learned that Kuhl's Mark IV SPECT could not collect data at high speed, he spent the next few years developing a high-speed tomographic device that could measure the xenon-133 clearance method. The completed xenon-133 tomography system was able to measure cerebral blood flow and brain activity not only in normal subjects, but also in patients with various cerebral ischemic diseases, including stroke. The tomographic images of cerebral blood flow provided important diagnostic information for areas deep in the brain. This type of information could not be obtained by measuring cerebral blood flow on the surface of the brain, demonstrating the importance of tomography (see Sect. 4.4.1).

Michel Ter-Pogossian and Michael Phelps at Washington University in St. Louis developed PETT, which was a full-scale PET device with detectors arranged in a hexagonal shape (Ter-Pogossian et al. 1975). They progressively increased the number of detectors and improved the image resolution and image reconstruction theory. Later, these two physicists had a disagreement and so went their separate ways in PET development. Ter-Pogossian began to develop a TOF-PET system that used time-of-flight (TOF) information from gamma rays (photons) to achieve high-speed acquisition. Phelps, on the other hand, developed a high-resolution PET system and started a venture company to contribute to the widespread use of PET systems. In addition to those two, at the Montreal Neurological Institute (MNI) Lucas Yamamoto (previously at BNL) developed a full-scale ring PET system dedicated to the head. At the Karolinska Institute in Stockholm, Lars Eriksson developed a ring-shaped PET system for the whole body. In Japan, Eiichi Tanaka and colleagues, at the National Institute of Radiological Sciences (NIRS), and Kazuo Uemura and Iwao Kanno, at the Akita Cerebrovascular Research Center (Akita Brain Institute), developed their own PET devices.

Column: Development of a Tomographic Device Inspired by Study Abroad

In 1978, I had the opportunity to study for six months with Niels Lassen in Copenhagen. The purpose of my study in Copenhagen was to develop an algorithm for calculating cerebral blood flow from xenon-133 brain clearance tomography data. At the time, the Copenhagen facility was developing a new high-speed SPECT system. Edda Sveinsdottir, who supported my work, was working on the development and design of that SPECT system. I had previously measured SPECT images of a patient rotating on a swivel chair in front of an Autofluoroscope. As I learned more about their work, I felt that if this small group in Denmark could develop a full-scale SPECT system, then a similar project should be possible for us in rural Japan. I immediately talked to my boss about this as soon as I returned to Japan. We decided that if we were going to develop a device we would need to develop a unique system. With the cooperation of Shimadzu Corporation, we began to develop a hybrid tomography device that could measure both SPECT and PET.

4.3.2 The Author's Challenge

In 1972, information was circulating that EMI in the UK had developed a computed tomography (CT) system using X-rays. The author thought that if tomography could be done with X-rays, then it should also be possible to do tomography with RIs. To confirm the algorithm for image reconstruction, the author first made a program that simulated filtered back-projection on the aforementioned 8-kiloword digital computer. The simulation took a couple of years to complete. The next step was to confirm the simulations with tomographic imaging of a RI phantom made from a Tupperware container. The phantom was placed on a table sitting on a broken cooling fan in front of a gamma camera called an Autofluoroscope. A Mabuchi miniature motor was then used to rotate the phantom in discrete angular steps. Tomographic images of the phantom were successfully obtained. In the mid-1970s, it was decided to make a system to measure human patients. An obsolete dental chair configured to rotate in discrete angular steps was used to rotate the patient in front of the Autofluoroscope (Fig. 4.14). This device measured the first human RI tomogram in Japan, which was a technetium-99m accumulated tomogram of a meningioma (Fig. 4.15, left) that agreed well with the X-ray CT image (Fig. 4.15, right).

Column: Kazuo Uemura, a Laissez-Faire Boss

Kazuo Uemura used to tell young radiologists that to understand brain pathology with angiography it is important to understand not only the anatomy of the blood vessels but also the cerebral blood flow. At a time before X-ray CT, the role of radiology was to identify diseased vessels with serial cerebral angiography. However, Uemura seemed to enjoy estimating cerebral blood flow by observing the flow of angiographic contrast medium from the arterial phase to the venous phase. He would place a two-inch detector over the temporal region, inject RI solution intravenously through the brachial vein, and calculate the mean transit time from the radioactivity curves of the left and right cerebral hemispheres. However, it was difficult to accurately estimate the cerebral blood flow because the input function is blurred as it traverses the heart and lungs. When he found Lassen's paper on the xenon-133 internal carotid artery injection method, he immediately made a six-channel xenon-133 detector to quantify cerebral blood flow. As a supervisor, Uemura gave his subordinates free reign. When the author, who graduated in electrical engineering, arrived at the Akita Brain Institute in 1970, Uemura asked the head of the Neurosurgery Department whether we could use their digital computer to analyze the xenon-133 brain clearance curves. We were like Sun Wukong playing on the palm of Buddha's hand. Uemura freely gave us ideas and toys to play with, but he did not nag us. He was unusually knowledgeable in science and engineering for a medical doctor, and when we started to develop a tomography device, Uemura understood its principles better than the author did.

Based on the success of the dental chair rotation system, the author and his colleagues were able to obtain the cooperation of Shimadzu Corporation and proceeded to develop a full-

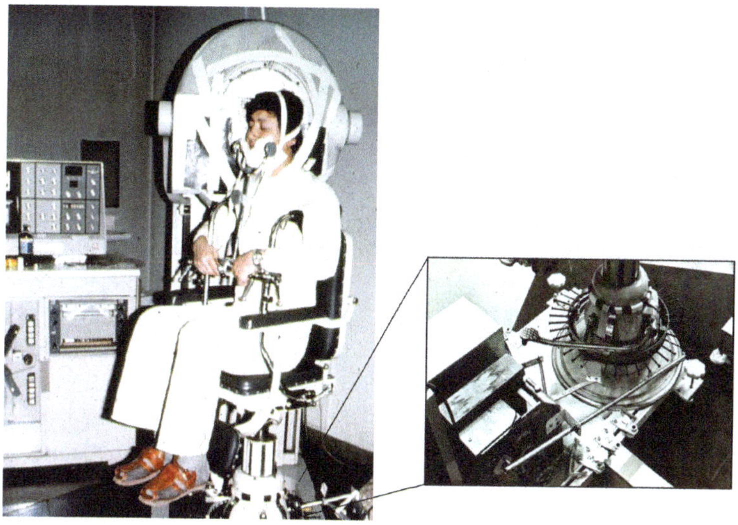

Fig. 4.14 Custom-made tomograph: The author and colleagues built a tomographic device using the parts of abandoned machines. A dental chair was rotated 360 degrees in 9-degree steps in front of an Autofluoroscope camera to measure brain projections. The tomogram was reconstructed from the projections using the filtered back-projection algorithm

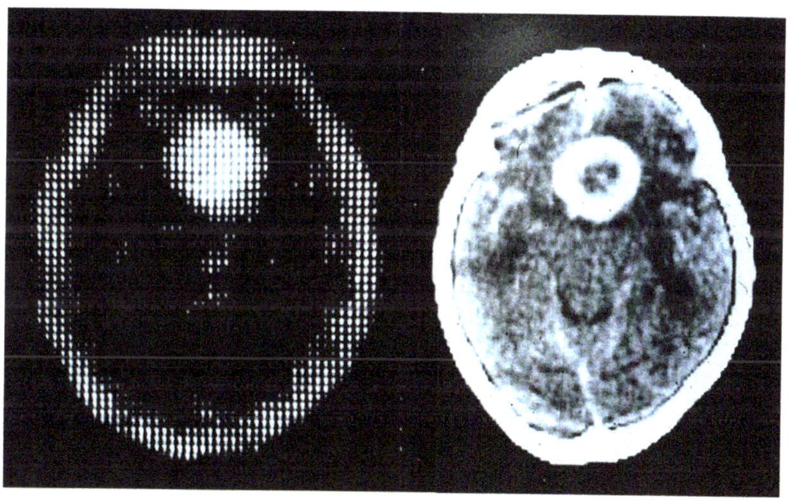

Fig. 4.15 The first tomogram of a patient measured with the custom-made SPECT: Accumulation of 99mTc pertechnetate due to BBB leakage within meningioma (tumor) was seen (left) and agreed with the region in the X-ray CT showing contrast media leakage (right)

scale tomography system. The members of the collaboration were young, ambitious, full of ideas and not afraid of failure, so they designed the world's first hybrid-type tomography system that could measure both PET and SPECT. The system was completed in 1980, and in the following year they succeeded in taking the first xenon-133 clearance tomographic images for cerebral blood flow measurements in Japan. The system was named HEADTOME, which is an abbreviation of "hybrid emission advanced dynamic tomography". Encouraged by this success, the collaboration continued to develop successor systems with higher resolution and performance. By the mid-1990s, several PET systems had been developed and a full-width at half-maximum of 6 mm reached, which was the world's best resolution for PET at the time (Fig. 4.16).

4.3.3 Demand for High-Speed SPECT

Although the development of tomography systems gained momentum, xenon-133 tomography systems required both high sensitivity and high scanning speed. Resolution and sensitivity are generally in conflict with each other, but the HEADTOME system design prioritized sensitivity at the expense of resolution. The need for high-speed dynamic images was not limited to the xenon-133 brain clearance method; high-speed imaging was also required for cardiac imaging. Therefore, a device that minimized mechanical motion during scanning was needed. The xenon-133 SPECT system developed in Copenhagen was like the early CT systems in that the entire gantry rotated at high speed while scanning linearly along each side of the square detector array.

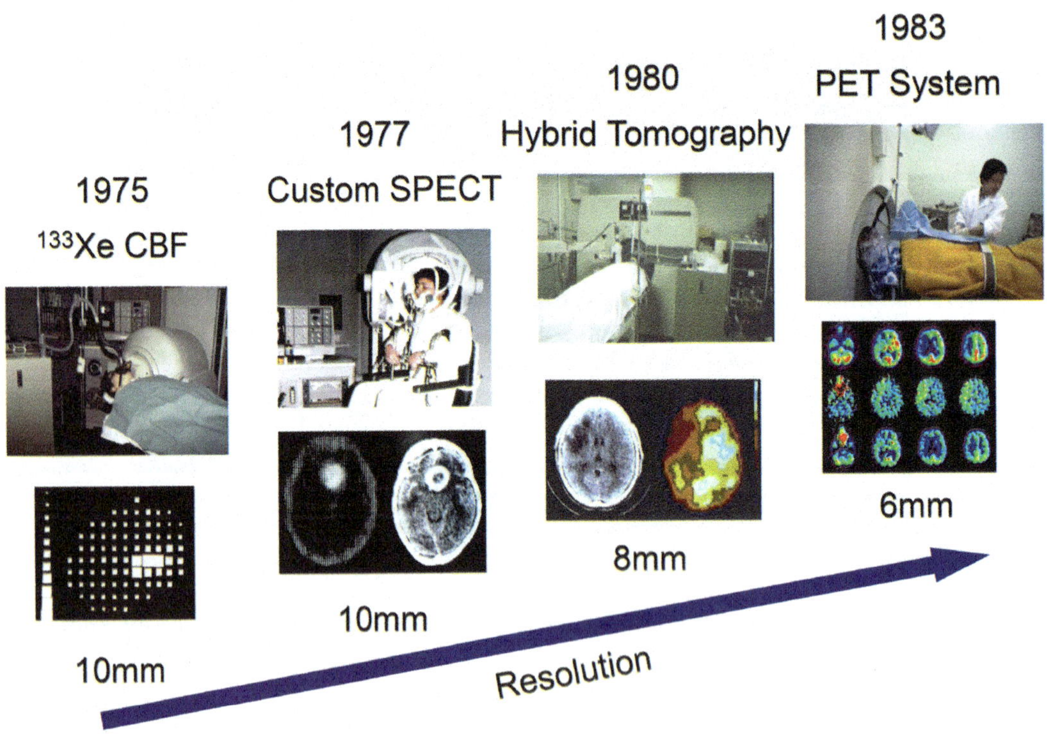

Fig. 4.16 Improvements in the resolution of measured cerebral blood flow: The authors developed devices for cerebral blood flow measurement at the Akita Brain Institute. Starting with a gamma camera that measured a non-tomographic map of the temporal hemisphere, the system evolved to a custom-made SPECT, then a hybrid tomography system, and eventually a PET system. Each of these developments improved the spatial resolution as well as the quantitative accuracy

In contrast, the SPECT system developed at the Akita Brain Institute was an ambitious and sophisticated design where the gantry was fixed and the detector array was in the shape of a ring, and data was collected by rotating a collimator ring. However, when chemical microsphere tracers labeled with technetium-99m became commercially available, the popularity of the xenon-133 cerebral clearance method declined. For the clinical diagnosis of cerebrovascular disease, high-resolution qualitative technetium-99m cerebral blood flow measurement was preferred over the quantitative xenon-133 dynamic method. As a result, brain-specific fast SPECT systems have disappeared. The role of quantitative cerebral blood flow measurement has shifted to iodine-123-labeled IMP (see Sect. 6.1.2) and PET measurements, as described below.

4.4 Noninvasiveness Is a Trend of the Times

It is well known that the functions of the brain are heterogeneously distributed throughout its complex structure. However, since Penfield's method, in which open skull surgery was performed to elucidate the brain function of epilepsy patients, cannot be used on healthy people, neurophysiologists had been searching for noninvasive methods to examine local cerebral blood flow.

4.4.1 Foresight to Introduce RIs

Sokoloff (Fig. 4.17) knew that, unlike Kety's laughing gas method, the use of RIs could solve the problem of measuring regional cerebral blood flow. Lassen was a pioneer in using noble gas RIs, which the author imagines was the result of discussions with Sokoloff and his colleagues at the National Institutes of Health (NIH). Sokoloff himself developed a method to measure glucose metabolism in animal brains using beta-emitting carbon-14-labeled deoxyglucose (DG) (Fig. 4.18). Specifically, he developed a quantitative autoradiography (ARG) method in which brain sections were taken a certain time after carbon-14-labeled DG administration and placed on a photosensitive film (see Sect. 6.4.2). The photosensitive film was exposed to the beta-rays emitted by the carbon-14, resulting in images with excellent resolution (Sokoloff et al. 1977). A few years later, Tatsuo Ido and colleagues at BNL developed fluorodeoxyglucose (FDG), which was labeled with positron-emitting fluorine-18 instead of carbon-14 (Ido et al. 1978). This breakthrough enabled PET measurement of glucose metabolism in human brain. The development of FDG was an extremely important step towards the current PET era.

Fig. 4.17 Louis Sokoloff (1921–2015)

Column: Sokoloff, a Giant in Brain Science

Sokoloff was the last of the four giants, Kety, Sokoloff, Ingvar, and Lassen, who pioneered the methods to measure cerebral circulation and metabolism. Sokoloff published his method for quantitative measurement of cerebral glucose metabolism in 1977. The author thinks it unfortunate that he did not receive the Nobel Prize later in his life. At meetings, he enthusiastically explained the theory and method of measuring brain glucose metabolism with the carbon-14-labeled DG method. In 1986, when Fox & Raichle presented their result that cerebral oxygen metabolism was lower than cerebral blood flow during neural activation, Sokoloff immediately argued that the PET measurement model was wrong. It took several years for him to come around, but he eventually accepted that oxygen metabolism is low during the activation state due to anaerobic metabolism, which is different from the resting state.

Sokoloff was a friendly, spirited old man. The last time the author talked with Sokoloff was at the IAEA conference in Strasbourg, France, in October 2000. When the author walked with him from the hotel to the conference hall, he mentioned that Seymour (Kety) had passed away, Niels (Lassen) had passed away, and David (Ingvar) had recently passed away, and now Louis (Sokoloff) is next. He seemed to enjoy the gossip on when he would die. Sokoloff passed away 15 years later at the age of 93 in 2015.

4.4.2 Definition of Invasiveness Changing with Time

From the perspective of Penfield's time, any method that does not directly touch the brain can be said to be noninvasive. Therefore, the laughing gas inhalation method and the xenon-133 brain clearance method were also noninvasive methods because they were performed with the brain intact. Since these methods involve direct puncture of the jugular vein or the internal carotid artery, they are now considered invasive and so are not used. Later, in the early PET era, arterial blood sampling from the radial artery or brachial artery was used to obtain an accurate measurement of arterial tracer concentration, which was used as an input function for quantitative measurement of dynamic tracer kinetics. Initially, PET was also treated as a noninvasive technique because it was less invasive than puncturing the proximal neck, as was required for the laughing gas method, or xenon-133 intracarotid arterial injection. However, with the passage of time,

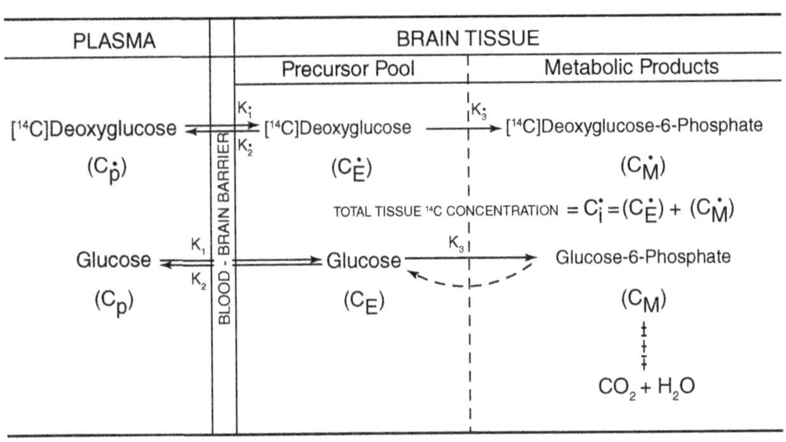

Fig. 4.18 Analysis model for quantifying glucose metabolism using 14C deoxyglucose (DG): Carbon-14-labeled DG (top row) and glucose (bottom row) are 6-phosphorylated in the brain. The glucose is metabolized to water and CO_2, while the carbon-14-labeled DG does not react after phosphorylation and stays there. (Sokoloff L, et al. J Neurochem. 1977)

arterial puncture itself has become highly invasive and, since 2000, PET that requires arterial blood sampling has been positioned as an invasive measurement. Recently, even the administration of RI-labeled tracers has been defined as invasive because of the exposure to radiation.

There are many advantages to having clinical examinations that are noninvasive. First, it lowers the hurdle for patient examination and thus encourages follow-up tests. Cerebral blood flow tests are important for determining the condition of the patient by tracking the changes in blood flow as the disease or other condition progresses. This is easier to do if the examination is noninvasive. Noninvasiveness is an important concept for high-quality medical examination and treatment.

4.4.3 PET and SPECT Are Now Standard Tools

Today, almost all nuclear medical imaging uses tomography. SPECT uses gamma-ray emitting RIs such as xenon-133, technetium-99m, and iodine-123, and PET uses positron emitting RIs. Some of the differences between SPECT and PET are briefly discussed below.

PET has much higher resolution and sensitivity than SPECT. The RI in PET emits positrons that annihilate with electrons in the immediate vicinity to produce a pair of 511 keV annihilation gamma-rays (equivalent to the masses of the positrons and electrons) traveling in opposite (180-degree) directions. In contrast, the xenon-133, technetium-99m, and iodine-123 RIs used in SPECT emit single-photon gamma rays that are collimated to estimate the gamma-ray source position and measure the tomogram. The collimator narrows the incidence angle of the detected gamma-rays, resulting in lower sensitivity. In addition, the resolution depends on the distance of the detector from the gamma-ray source. Another advantage that PET has over SPECT is that absorption/attenuation correction and scattering correction, which are essential for quantitative radiation measurement, are more straightforward (Fig. 4.19 & Table 4.1).

It is possible to manufacture radiopharmaceutical tracers that mimic the compounds of living organisms. Various tracers labeled with carbon-11 and fluorine-18, which is a RI analog of hydrogen, can be synthesized for PET. On the other hand, although iodine-123 can be readily used as a SPECT RI label because it is a halogen group element with high chemical reactivity, technetium-99m must be chelated before it can be used as a label in studies of living organisms.

Table 4.1 SPECT and PET

System	SPECT	PET
Photon detection	Collimator	Coincidence
Scatter correction	Model	Measure
Attenuation correction	Model	Measure
Resolution	~1 cm	0.3 cm
Axial FOV	>40 cm	<25 cm
In plane FOV	>30 cm	<40 cm

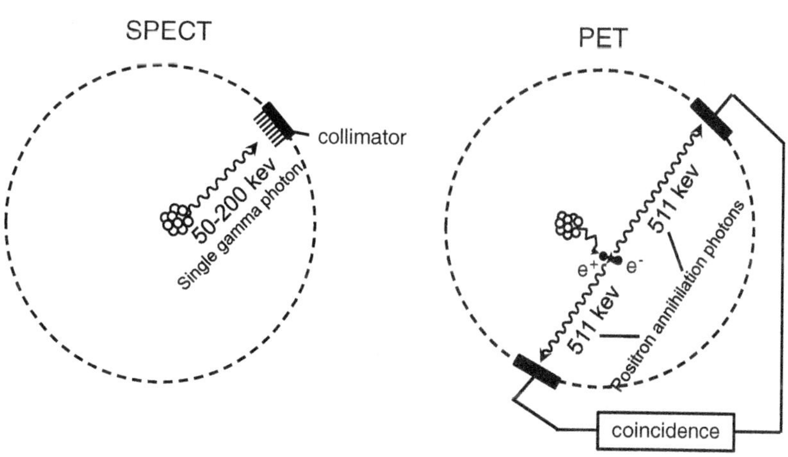

Fig. 4.19 Principles of SPECT and PET: (Left) SPECT detects single-photon gamma rays using a collimator. (Right) PET detects positron annihilation gamma-rays using opposing detectors and a coincidence circuit. In both systems, many detectors are arranged in a ring form

4.4.4 PET Opens the Door to a New Era

From 2000, the development and production of PET systems began to be dominated by European and American companies. Detectors have evolved from single crystals to multicrystals, and photoelectric converters have progressed from vacuum tubes to semiconductors. Even though the most common clinical application of PET is the screening and diagnosis of malignant tumors, its capabilities are much broader.

Neural transmission/reception, metabolic function and the synthesis of abnormal proteins can now be observed at a high precision. It is now possible to image with a field-of-view that is wide enough to cover the entire brain. Recently, Simon Cherry, Terry Jones and colleagues built an ultra-high-sensitivity total-body PET system that can measure the total-body behavior of the tracer after a single administration (Badawi et al. 2019). This will enable "systemic medicine", where the body is treated as a whole system, rather than the current organ-based medicine where doctors treat organs or body regions separately. For example, even though serotonin is an important neurotransmitter in the brain, it is actually distributed throughout the whole body. The brain contains only a few percent of the total amount of serotonin in the body. Dementia is also thought to be caused by abnormal protein accumulation, and abnormal accumulation in the brain may be linked to abnormal accumulation in other parts of the body. For diseases like these, a total-body PET scanner would be able to observe the distribution throughout the whole body in a single dynamic measurement after a single tracer dose. It is expected that total-body PET will lead to a breakthrough in medical care where all diseases are treated as systemic diseases.

References

Badawi RD, Shi H, et al. First human imaging studies with the Explorer total-body PET scanner. J Nucl Med. 2019;60(2):299–303.

Fick A. The development of heat by muscular activity. Science. 1881;2(80):618–21.

Hamilton, et al. Further analysis of the injection method, and of changes in hemodynamics under physiological and pathological conditions. Am J Physiol. 1931;99:534–51.

Ido T, et al. Labeled 2-deoxy-D-glucose analogs. 18F-labeled 2-deoxy-2-fluoro-D-glucose, 2-deoxy-2- fluoro-D-mannose and 14C-2-deoxy-12-fluoro-D-glucose. J Label Compds Radiopharm. 1978;24:174–83.

Kanno I, Uemura K, et al. HEADTOME: a hybrid emission tomograph for single photon and positron emission imaging of the brain. J Comput Assist Tomog. 1981;5(2):216–26.

Kety SS, Schmidt CF. The determination of cerebral blood flow in man by the use of nitrous oxide in low concentrations. Am J Physiol. 1945;143:53–66.

Kety SS, Schmidt CF. The nitrous oxide method for the quantitative determination of cerebral blood flow in man: theory, procedure and normal values. J Clin Invest. 1948;27(4):476–83.

Kuhl DE, Edwards RQ, et al. The Mark IV system for radionuclide computed tomography of the brain. Radiology. 1976;121(2):405–13.

Lassen NA, Hoedt-Rasmussen K. Human cerebral blood flow measured by two inert gas techniques. Comparison of the Kety-Schmidt method and the intra-arterial injection method. Circ Res. 1966;19(4):681–94.

Lassen NA, Ingvar DH. The blood flow of the cerebral cortex determined by radioactive krypton. Experientia. 1961;15(17):42–3.

Mosso A. Sulla circolazione del cervello dell'uomo. Atti della R. Accademia dei Lincei. 1880;5:237–358.

Roy CS, Sherrington CS. On the regulation of the blood-supply of the brain. J Physiol. 1890;11(1–2):85–158, 17.

Schlant RC, et al. Determination of central blood volume; comparison of Stewart-Hamilton method with direct measurements in dogs. Am J Physiol. 1959;196(3):499–501.

Sokoloff L, Reivich M, et al. The [14C]-deoxyglucose method for the measurement of local cerebral glucose utilization: theory, procedure, and normal values in the conscious and anesthetized albino rat. J Neurochem. 1977;28(5):897–9.

Stewart GN. The pulmonary circulation time, the quantity of blood on the lungs and the output of the heart. Am J Physiol. 1921;58:20–44.

Ter-Pogossian MM, Phelps ME, et al. A positron-emission transaxial tomograph for nuclear imaging (PETT). Radiology. 1975;114(1):89–98.

Zierler KL. Equations for measuring blood flow by external monitoring of radioisotopes. Circ Res. 1965;16:309–21.

Measuring Cerebral Blood Flow: Theory

5

Abstract

This chapter describes the theory for quantitative measurement of cerebral blood flow. Cerebral blood flow is perfusion flow that can be measured using radioisotope-labeled tracers. There are two types of tracers, transit-type and plug-type. The transit-type tracers (diffusible tracers and nondiffusible tracers) travel through the vessels and the mean transit time (MTT) is used for cerebral blood flow calculation. The plug-type tracers (microsphere tracers and chemical microsphere tracers) are trapped in the vessels before they can transit, and the density of the plugged tracer is used for cerebral blood flow calculation. Accurate calculation needs to consider the partition coefficient and first-pass extraction. In tracer kinetics, a compartment model and an accurate input function are necessary. The kinetic model is applied to oxygen-15 gas studies, FDG studies, neurotransmission ligand analysis, and various other tracers. When analyzing cerebral blood flow, it is possible to use either the absolute or relative value.

Keywords

Diffusible tracer · Nondiffusible tracer · Microsphere tracer · Chemical microsphere tracer · Partition coefficient · First-pass extraction · Compartment model · Input function

5.1 Bulk Flow Versus Perfusion Flow

There are two definitions for blood flow. Bulk flow is like the flow of bulk water in a fixed channel or river. On the other hand, perfusion flow is the way that water flows through a medium like sand or a shrub (Table 5.1) (Fig. 5.1).

5.1.1 Units of Bulk Flow and Perfusion Flow

The units used to describe measurements of bulk flow and perfusion flow are different. Bulk flow is defined as the flow volume (mL) per unit time (min); therefore, mL/min. On the other hand, the unit of perfusion flow is defined as the flow volume (mL) per unit time (min) through a unit of tissue mass (g); therefore mL/min/g. Since the specific weight of tissue equals unity, the flow volume and the tissue mass cancel, so the unit of perfusion flow can simply be defined as the reciprocal of time. This definition is often useful when measuring cerebral blood flow. Bulk flow can be measured directly by, for example, attaching an electromagnetic flowmeter that wraps around the blood vessel. Perfusion flow, however, is measured only indirectly using tracers.

I. Kanno, *What Cerebral Blood Flow Tells Us*, https://doi.org/10.1007/978-981-97-6382-5_5

5.1.2 Measurement of Bulk Flow

Bulk flow can be measured by imaging blood flow in single vessels. Since macroscopic bulk flow follows the morphology, or path of the blood vessel, it can be called morphological blood flow. For example, angiography, which is performed using CT or X-rays after injection of an intravascular iodine contrast agent, is used to image the vessel morphology. Magnetic resonance angiography (MRA) can also be used to image blood vessel morphology after the injection of gadolinium, which is a paramagnetic contrast agent. These angiographic techniques provide important anatomical information for the diagnosis of vascular disease. Furthermore, if the input function of the contrast agent is known, the cerebral blood flow can be estimated from the kinetics of the contrast medium in the angiographic data.

In rodent experiments, after thinning the skull and attaching a closed cranial window, two-photon-excitation laser-scanning fluorescence microscopy (hereafter "two-photon microscopy") enables microscopic evaluation of vessels in the cerebral cortex (see Sect. 6.4.6). By using different fluorescent dyes as tracers, it is also possible to independently observe red blood cells and plasma. Vessel diameter can be estimated from the width of the plasma flow. It is also possible to directly measure blood vessel diameter using genetically-modified Tie2 mice that express green fluorescent protein (GFP) in their endothelial cells. In this way it is possible to precisely evaluate the relationship between blood vessel diameter and blood flow velocity.

Table 5.1 Bulk Flow and Perfusion Flow

Definition	Bulk flow	Perfusion flow
Flow path	Fixed path	Uncertain bush
Role	Locomotion	Exchange
Meaning	Morphologic flow	Functional flow
Unit	mL/min	mL/min/g

5.1.3 Measurement of Perfusion Flow

Unlike the case for bulk flow, direct measurements of perfusion flow are not possible and do not provide morphological information. Cerebral blood flow is essentially the measurement of perfusion flow through a tissue mass with a capillary network. Perfusion flow is measured using a

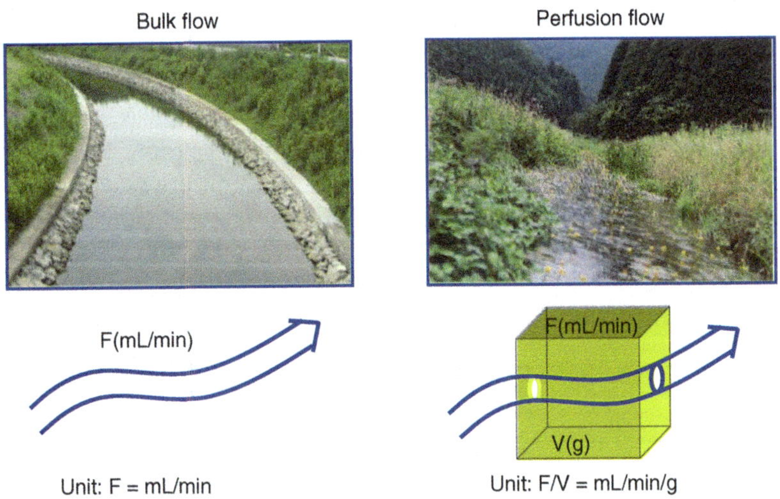

Fig. 5.1 Bulk flow and perfusion flow: Bulk flow is like the flow of water along a fixed passage (left), while perfusion flow is like the flow of water seeping through sand or shrubbery (right). The units of measurement are different for each type of flow

tracer that disperses into the capillary network through the bloodstream.

Since the fundamental role of cerebral blood flow is to supply nutrients to brain tissue via the capillary network, the target of cerebral blood flow measurements is perfusion flow rather than bulk flow. In this book, unless otherwise stated, cerebral blood flow refers to perfusion flow. As described in Chap. 4, the first quantitative measurement of cerebral blood flow was performed by Kety using N_2O as a tracer. Lassen improved on this by using RI noble gases, which are less invasive, have no anesthesia effect and enable regional measurement of blood flow. N_2O and RI noble gases are tracers that diffuse uniformly into brain tissue from the vasculature. There are now also other types of tracers that plug the capillary network rather than traveling right through. The principle for estimating cerebral blood flow for tracers of that type will be explained in detail later in this chapter (see Sects. 5.3.1, 5.3.2, 5.3.3, 5.3.4, and 5.3.5).

5.1.4 From Bulk Flow to Perfusion Flow

Microscopic measurements of bulk flow (mL/min) can be used to indirectly estimate perfusion (mL/min/g). For example, two-photon microscopy allows us to measure the bulk flow in the capillaries passing through a single thin cross-section. By integrating the bulk flow through several adjacent cross-sections, the total bulk flow (in mL/min) through the integrated brain tissue volume can be measured. Dividing the total bulk flow by the mass of the brain tissue volume provides an estimate of the perfusion flow (in mL/min/g) through that volume. That is, it is possible to quantitatively estimate the perfusion of a three-dimensional volume by integrating the bulk flow in the Z direction through thin two-dimensional cross-sections in the X-Y plane. At the same time, the mechanisms regulating blood vessel diameter, red blood cell velocity and plasma velocity, which are not seen in macroscopic perfusion measurements, can be analyzed for research into the mechanisms of cerebral blood flow regulation (Kanno and Masamoto 2016). Furthermore, with the development of genetic engineering to label specific cells and molecules, it has become possible to use genetically-engineered mice with fluorescent proteins expressed in specific target cells, making it possible to not only observe changes in the dilation and constriction of arterioles, capillaries, and venules in real time, but also the so-called neurovascular unit (NVU) (see Sect. 3.3.2) as a whole.

5.2 Tracers Used for Blood Flow Measurement

This section describes tracers that are indispensable when measuring cerebral blood flow. As mentioned in the previous section, tracers enable indirect measurement of cerebral blood flow by labeling the blood and observing its flow. Tracers are like the rings attached to the legs of migratory birds to track their behavior. Counting (or measuring) the number of birds wearing the leg rings at a given site allows us to indirectly estimate the overall behavior of the birds. Similarly, blood flow through individual capillaries cannot be measured, but by labeling the blood with a tracer and measuring the tracer concentration it is possible to indirectly measure blood flow. RI tracers can be used not only to measure cerebral blood flow but also to observe the route through the target organ after being administered into the bloodstream. From this information it is possible to evaluate biological function with the aid of a mathematical model.

There are four types of tracers for measuring cerebral blood flow: (a) diffusible tracers that freely pass through the blood–brain barrier (BBB) from the capillaries into brain tissue, (b) nondiffusible tracers that cannot pass through the BBB and remain inside the blood vessels, (c) microsphere tracers that plug the capillaries because they are slightly larger than the capillary diameter. In addition, there are (d) chemical microsphere tracers that are lipophilic in the bloodstream, but after passing through the BBB into brain tissue they are hydrolyzed by enzymes and remain in the perivascular area as they are unable to return into the vessel. Chemical microsphere tracers were developed in the 1990s and from around the year 2000 they have become widely commercialized as SPECT tracers. The following briefly explains each of the four tracer types.

5.2.1 Diffusible Tracers

Diffusible tracers can freely enter and exit the brain tissue through the BBB of cerebral blood vessels. As soon as a diffusible tracer arrives at a capillary, it diffuses into the surrounding brain tissue until the tracer concentrations in the intravascular blood and extravascular brain tissue reach equilibrium (Fig. 5.2). Capillaries are densely distributed in normal tissue, so diffusible tracers are usually uniformly distributed throughout the brain tissue. However, the solubility of the tracer in blood is different to that in brain tissue, which means that the tracer concentrations in blood and brain tissue are different. The ratio of tissue concentration to blood concentration is called the brain-blood partition coefficient (see Sect. 5.3.1). When a diffusible tracer is quickly administered, it is assumed that the tissue-to-blood tracer concentration at the capillaries instantaneously reaches equilibrium and the

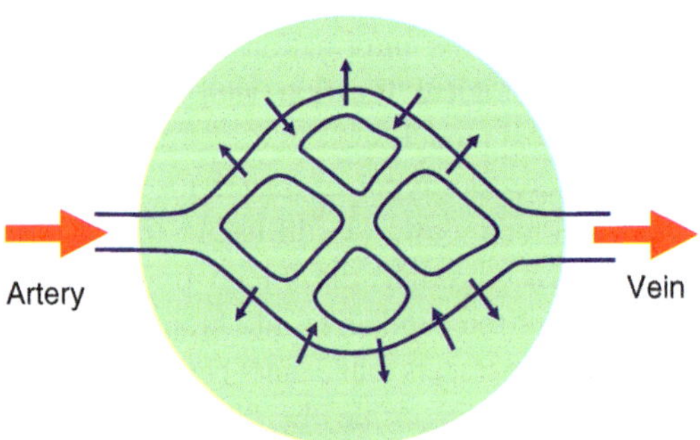

Fig. 5.2 Diffusible tracer: A diffusible tracer can diffuse freely between the capillaries and brain tissue. The distribution volume of these tracers is equal to the whole brain. The cerebral blood flow is calculated based on the mean transit time of the tracer through the whole brain tissue

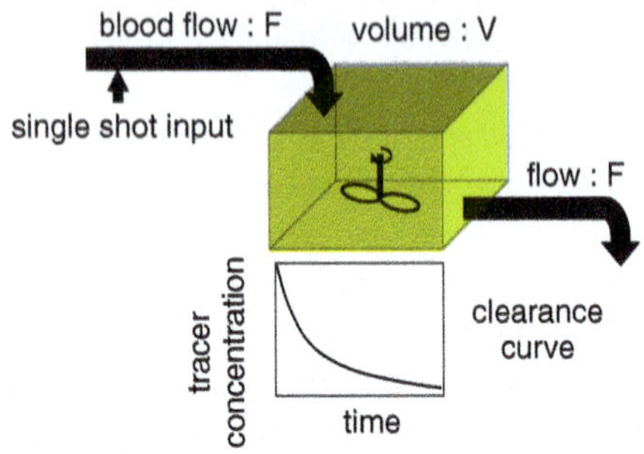

Fig. 5.3 Water tank model: A system with a volume *V* and a flow *F*. When the tracer is instantaneously administered at the inlet, there is an initial peak in tracer concentration followed by exponential decay with time

tracer is thereafter flushed out by fresh blood flow. Measurement techniques based on this principle are called "clearance methods".

The analysis of clearance method data is based on the aquarium model where the entire brain is thought of as an aquarium (Fig. 5.3). This model assumes that the aquarium has a water inlet and outlet and is equipped with a stirring device to instantly mix incoming water. The volume of the aquarium is *V*. As it is also assumed that the inflow and outflow of water are the same, the water volume is constant. If ink is dropped into the inlet of the aquarium, the concentration of ink in the aquarium will be maximum at the moment it is dropped, and from then on the ink concentration is diluted in an exponential manner by the inflow of new water. The decay constant of the exponential function is proportional to the inflow speed of the water and inversely proportional to the volume of the aquarium. If the inflow rate is *F* (mL/min), the volume is *V* (mL), and the initial concentration of the ink is C_0, the clearance function of the ink concentration at time *t*, *C* (t), is

$$C(t) = C_0 \times \exp\{-(F/V)t\}.$$

Before PET, diffusible tracers used for clinical cerebral blood flow measurement included inert RI gases such as xenon-133 and krypton-85, while oxygen-15-labeled water is widely used in PET. Diffusible tracers have a larger mean transit time (MTT) than nondiffusible tracers because the effective volume of the tracer is almost equivalent to the volume of the brain tissue. From the standpoint of radioactivity measurements, the advantage of measuring tracer concentration on a slower time-scale is that it enhances the accuracy of cerebral blood flow measurements. Slower measurements prolong the sampling interval, which provides a higher signal-to-noise ratio and subsequently allows more accurate estimates of the cerebral blood flow.

Iodine-123-labeled IMP, which was initially treated as a chemical microsphere tracer but is now treated as a diffusible tracer, can quantitatively measure cerebral blood flow because it is chemically inert in brain tissue and returns directly into blood vessels. However, because the brain-blood partition coefficient of iodine-123-labeled IMP is very large, it takes a very long time for the brain tissue concentration and blood concentration to reach equilibrium. The tracer concentration of the brain tissue can therefore be measured slowly, which results in a higher signal-to-noise.

5.2.2 Nondiffusible Tracers

Nondiffusible tracers flow through the inside of blood vessels without diffusing out into the brain tissue. The administered tracer flows directly downstream within the blood vessels (Fig. 5.4). Therefore, the volume through which a nondiffusible tracer passes is extremely small compared to that of a diffusible tracer. With respect to the aquarium model, the volume of the aquarium for nondiffusible tracers is extremely smaller than diffusible tracers (Fig. 5.5).

As the measurement sensitivity and speed of CT and MRI have improved, it is now possible to measure changes in the concentration of angiographic contrast agents, and from this information the mean transit time (MTT) can be accurately estimated when the contrast agent is considered as a tracer. In humans, the volume of cerebral blood vessels is 3–5% of that of the total brain, so the time for the tracer to pass through the brain (~2–10 s) is 3–5% of the time it takes

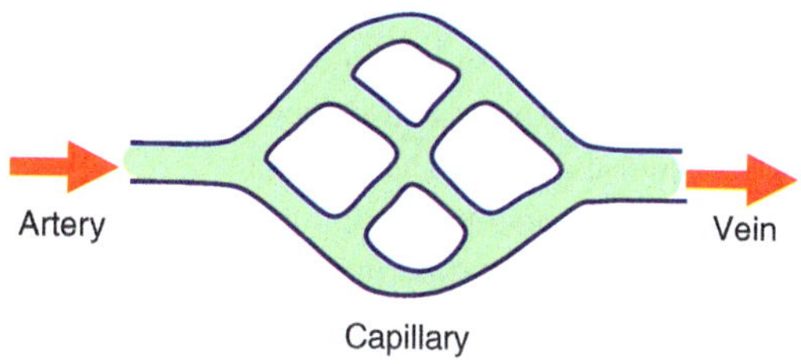

Fig. 5.4 Nondiffusible tracer: This tracers cannot freely diffuse out of the vessel and remain inside the vessel lumen. The distribution volume is equivalent to the inside of the vascular space. Cerebral blood flow is calculated based on the mean transit time of the nondiffusible tracer through the vascular space

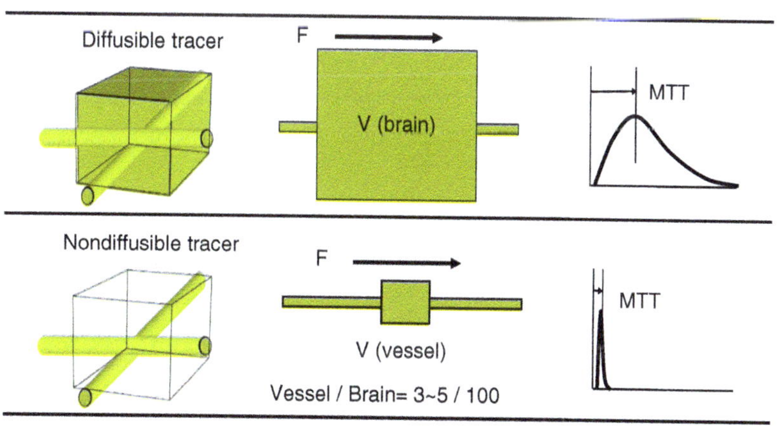

Fig. 5.5 Relationship between flow *F*, volume *V*, and mean transit time *MTT*: Mean transit time *MTT* is proportional to the volume *V* that the tracer passes through and inversely proportional to flow rate *F* ($MTT = V / F$). In the brain, *V* differs by two orders of magnitude for diffusible and nondiffusible tracers, so the *MTT* also differs by two orders of magnitude

for a diffusible tracer. This is approximately the same as the MTT of the blood itself. As a result, large errors can occur when the transit time of the blood is disrupted by the formation of abnormal vasculature in cerebrovascular deficiency.

5.2.3 Microsphere Tracers

Microsphere tracers are exclusively used in animal experiments and consist of particles that have a diameter larger than the capillary inner diameter so that they plug the vessel (Fig. 5.6). Cerebral blood flow is estimated from the number of captured microspheres. When using a microsphere tracer in practice, the microsphere tracer is injected into the left ventricular cavity of the animal. The tracer will be uniformly diluted in the blood flowing through the aorta and is thereafter distributed throughout the body. The tracer flows toward the periphery, bifurcating through the arteries to plug the capillaries in proportion to the blood flow. After tracer administration, the brain is removed, frozen and sliced, the density of microspheres lodged in the brain tissue is measured for each brain slice, and the cerebral blood flow is determined from the density of microspheres.

The author met Ter-Pogossian (also see Sect. 6.2.3), who was a pioneer of PET and cerebral blood flow measurement, and asked him which method had the best measurement accuracy. He immediately answered that it was the microsphere method, and clearly stated that microsphere tracers are the gold standard for cerebral blood flow measurement. Ter-Pogossian said that the microsphere method was the most accurate. He continued by saying that diffusible tracers require the tracer concentration in time, the input function of the tracer, the brain-blood partition coefficient, and several assumptions to accurately estimate cerebral blood flow. However, microsphere tracers only require direct counting of the number of particles and no assumptions are needed. Nevertheless, the author was overenthusiastic about a new algorithm that had been developed for tomographic cerebral blood flow measurement using a diffusible tracer (Kanno and Lassen 1979). At that time the author had no experience in animal experiments, so it took him some years to understand the simplicity of the microsphere method. The author gratefully remembers Ter-Pogossian's sincere answer to the impertinent question of a naïve young researcher.

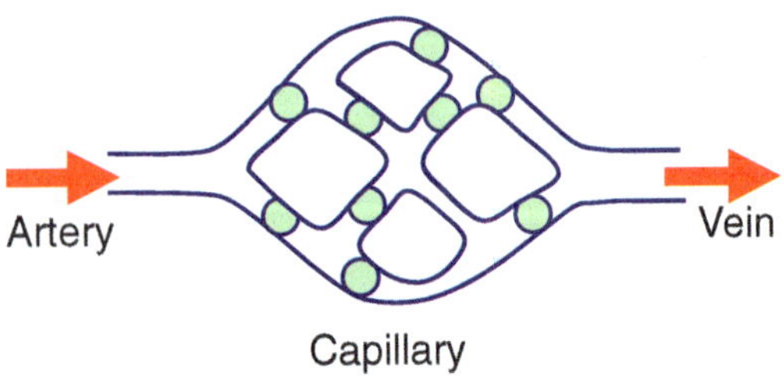

Fig. 5.6 Microsphere tracer: When a microsphere tracer reaches the capillaries, those microspheres that are larger than the capillary diameter plug the capillary and stay there. The blood flow is calculated by counting the number of microspheres plugging the vasculature

5.2.4 Chemical Microsphere Tracers

In the late 1980s, a breakthrough occurred in the development of tracers for cerebral blood flow measurement. A completely new type of microsphere tracer was developed for SPECT that could be applied for clinical use in humans. These were cleverly designed biochemical tracers that are trapped in the vessels. We call them chemical microsphere tracers. The tracers are designed to be lipophilic so that they penetrate the BBB and enter the tissue surrounding the blood vessel. However, once outside the blood vessel, enzymes in the brain tissue change the tracer to a water-soluble (hydrophilic) substance through hydrolysis and other reactions. As a result, the tracer cannot penetrate the blood vessel wall in the opposite direction to return into the blood vessel (Fig. 5.7). The chemical microsphere tracer remains in the brain tissue around the capillary and is captured there. The tracer remains distributed around the capillaries in proportion to the cerebral blood flow in the same way as microsphere tracers, and the cerebral blood flow is estimated from its density in the tissue (Fig. 5.8). Even though numerous chemical microsphere tracers were developed, only two of those, technetium-99m-labeled hexamethyl-propylene-amine-oxime (HM-PAO) and technetium-99m-labeled ethyl-cysteinate-dimer (ECD), are still in clinical use.

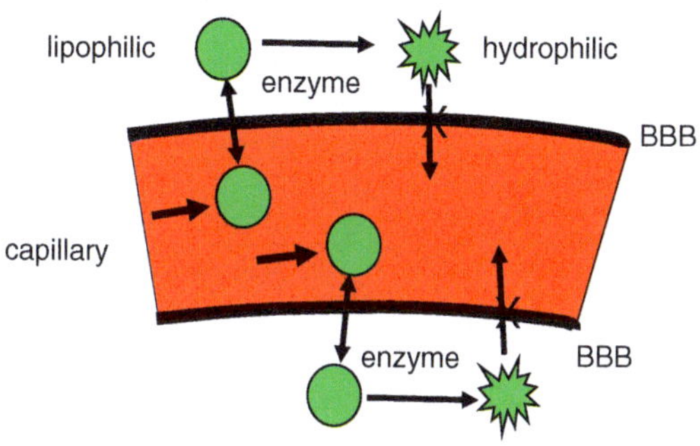

Fig. 5.7 Biochemical behavior of a chemical microsphere tracer: Chemical microsphere tracers diffuse out of the capillary through the BBB to the brain tissue. In the tissue they undergo hydrolysis by active enzymes and are changed to a hydrophilic form so that they cannot return to the blood vessels, so they remain in the brain tissue surrounding the capillaries

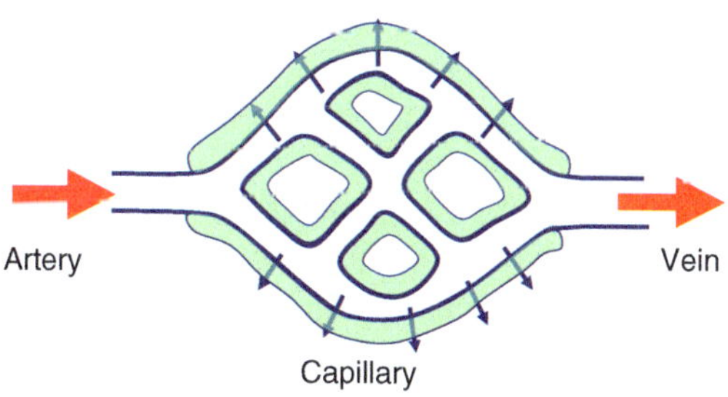

Fig. 5.8 Chemical microsphere tracer: Chemical microspheres diffuse biochemically out of the capillary and then remain in the brain tissue surrounding the capillaries. Cerebral blood flow is calculated based on the same principle as used for microsphere tracers

However, there are a few limitations when using chemical microsphere tracers. A portion of the tracer that reaches the capillaries passes through into the veins without exiting to the tissue. Also, a portion of the tracer that exits into the tissue will return to the blood vessels in a lipid-soluble state without undergoing enzymatic reactions. The returned portion leads to underestimation of the cerebral blood flow. The degree of underestimation will vary for different chemical microsphere tracers because the returned portion depends on the surrounding biochemical environment. In addition, the distribution of enzymes and enzyme activity varies depending on pathologies such as cerebral ischemia, so care must be taken when interpreting cerebral blood flow measured in pathological tissues.

With the widespread use of high-resolution SPECT systems, together with health insurance support (in Japan) for these tracers as cerebral blood flow measurement pharmaceuticals, many software applications have been developed to facilitate the clinical diagnosis of neurological disorders (e.g. cerebrovascular disease, psychiatric disorders and dementia) based on the distribution pattern of the qualitative regional cerebral blood flow. Later, central benzodiazepine receptor imaging agents, such as iodine-123-labeled Iomazenil, and dopamine transporter affinity imaging agents, such as iodine-123-labeled Ioflupane (FP-CIT), were also covered by health insurance. As a result, the clinical application of SPECT to degenerative diseases such as Parkinson's disease is becoming increasingly popular.

5.3 Calculating Cerebral Blood Flow

Methods for calculating cerebral blood flow can be divided into two categories depending on the tracer used. In the case of transit-type

Table 5.2 Calculation of Cerebral Blood Flow for the Various Tracer Types

Tracer type	Diffusible	Nondiffusible	MS[a]	CMS[b]
Input function	Need	Need	Not	Not
First pass extraction	Need	Need	Not	Need
Partition coefficient	Need	Need	Not	Not
Measurement speed	Slow	Fast	Slow	Slow
Calculation principal	Mean Transit Time		Distribution Density	

[a]*MS* microsphere
[b]*CMS* chemical microsphere

tracers, that is, diffusible tracers and nondiffusible tracers, the calculation is based on measuring the mean transit time (MTT) of the tracer through the area of interest. On the other hand, in the case of the plug-type tracers, that is, microsphere tracers and chemical microsphere tracers, the calculation is based on the density of the tracer plugging the measurement region (Table 5.2). At this point, it is important to incorporate the characteristics of each tracer into the calculation. Below is a brief explanation of these characteristics.

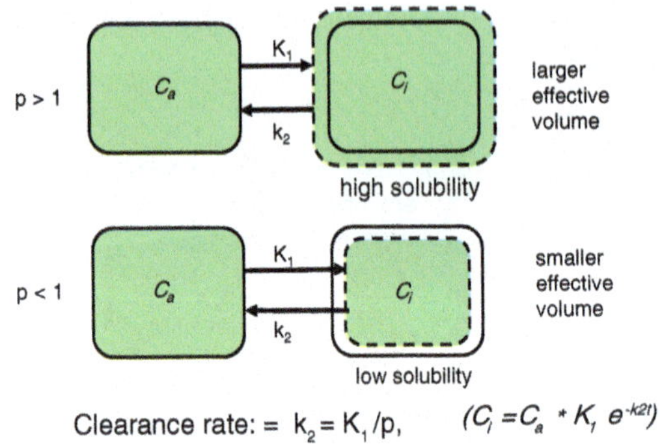

Fig. 5.9 Effective distribution volume and the partition coefficient: The partition coefficient (p) is defined as the ratio of a tracer's solubility in tissue to that in blood. It effectively corresponds to the expansion and contraction of the virtual tracer ~~distribution~~ volume in the tissue (dashed rectangle), which corresponds to effective changes in the rate constant of the tracer (see Sect. 5.5.1 for more detail)

5.3.1 Brain-Blood Partition Coefficient (*p*)

Diffusible tracers are chemically inert substances, but when they diffuse into the brain tissue the effective-volume of the tracer differs from the physical tissue volume depending on the solubility of the tracer in the tissue (physical solubility). In other words, the effective-volume of the tracer will be different from the physical volume. As described in Sect. 5.2.1, the mean transit time (MTT) is defined for the effective volume. When a diffusible tracer diffuses into brain tissue, it is assumed that the ratio of tracer solubility in blood and in brain tissue is the same as the ratio of tracer concentration in blood and in brain tissue at equilibrium, and this ratio is defined to be the partition coefficient (p) (Fig. 5.9). Explicitly, the partition coefficient (p) is the ratio of the tracer concentration in blood (mL^{-1}) to that in brain tissue (g^{-1}), so the accompanying unit is mL/g. If the tracer is more soluble in brain tissue than in blood, the partition coefficient is >1, and the effective-volume of the tracer is larger than the physical brain tissue volume (Fig. 5.9, upper panel). If the tracer is less soluble in brain tissue than in blood, the effective-volume will be smaller than the physical brain tissue volume so that $p < 1$ (Fig. 5.9, lower panel).

For example, the partition coefficients for gray matter and white matter in brain tissue are different due to their different physical properties. Since xenon-133 (which has a high atomic number and is heavy), is relatively lipid-soluble, its partition coefficient in white matter, which contains a large amount of lipid in the myelin that surrounds axons, has been experimentally determined to be $p_w = 1.5$. On the other hand, the partition coefficient of xenon-133 in gray matter, which contains less lipid, is much lower at $p_g = 0.8$. In contrast, the oxygen-15-labeled water used in PET has a partition coefficient of $p_w = 0.82$ for lipid-rich white matter and $p_g = 0.98$ for gray matter. However, since the spatial resolution of SPECT and PET is not high enough to resolve white matter and gray matter independently, the actual analytical model assumes a white-to-gray matter ratio of 50:50 and uses the average value of the gray and white partition coefficients as the partition coefficient of brain tissue. In other words, the partition coefficient used in practice for xenon-133 is $p = 1.15$ (Hoedt-Rasmussen et al. 1966) and that for oxygen-15-labeled water is $p = 0.9$ (Herscovitch and Raichle 1985).

In general, RI tracers require a complex compartmental model analysis (see Sect. 5.5) with the brain tissue divided into multiple compartments. The brain-blood partition coefficient is a fundamental characteristic of each tracer used to measure brain function. When a tracer moves from the blood compartment to the various compartments of brain tissue, the average tracer concentration in the brain tissue and the concentration in the blood are still expressed as a ratio to the concentration in the blood. In other words, these can be defined as an effective distribution volume.

5.3.2 First Pass Extraction (*E*)

All tracers are administered into the arteries, pass through the capillaries and then flow into the veins. The fraction of the tracer that passes from the vascular lumen into the tissue in a single passage is called the first-pass extraction (E). Ideally, 100% of a diffusible tracer would go out of the vessel into the tissue (i.e. complete extraction), so the extraction should be $E = 1$. However, in practice some portion of the diffusible tracer remains inside the vessels so in reality $E < 1$. On the other hand, an ideal nondiffusible tracer should remain 100% inside the vessels (i.e. no extraction) so $E = 0$, but in reality a small portion will enter the tissue, making $E > 0$ (Fig. 5.10). In addition, the first-pass extraction, E, is a function

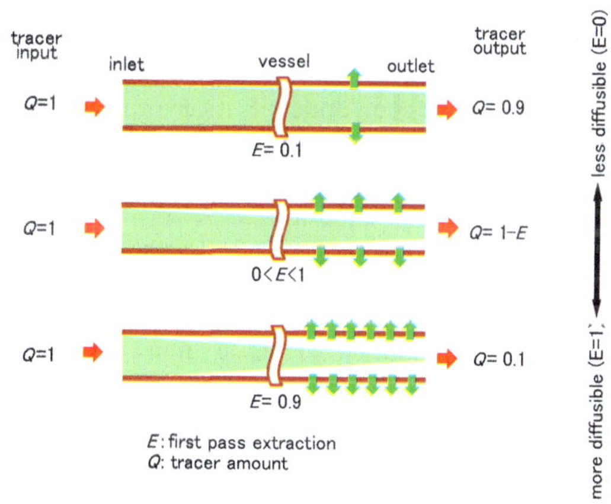

Fig. 5.10 First pass extraction: The first pass extraction (E) is defined as the fraction of tracer extracted from the vessel during the first passage. If the tracer is perfectly diffusible then $E = 1$, and if it is perfectly non-diffusible then $E = 0$. The value of E for an actual tracer is $0 < E < 1$

of the cerebral blood flow (CBF) as in the Renkin-Crone equation (Renkin 1953, Phelps et al. 1981)

$$E = 1 - \exp(-\mathrm{PS}/\mathrm{CBF}),$$

where *PS* is the product of the vessel permeability (*P*) and vessel surface area (*S*). This equation means that the first-pass extraction depends on the blood flow. That is, the higher the blood flow, the smaller the first-pass extraction. The amount of tracer taken up by brain tissue also varies in a nonlinear manner, so when a chemical microsphere tracer is used, its extraction into the brain is not directly proportional to the cerebral blood flow, which results in an underestimation of the actual cerebral blood flow in measurements (Fig. 5.11) (Iida et al. 1996).

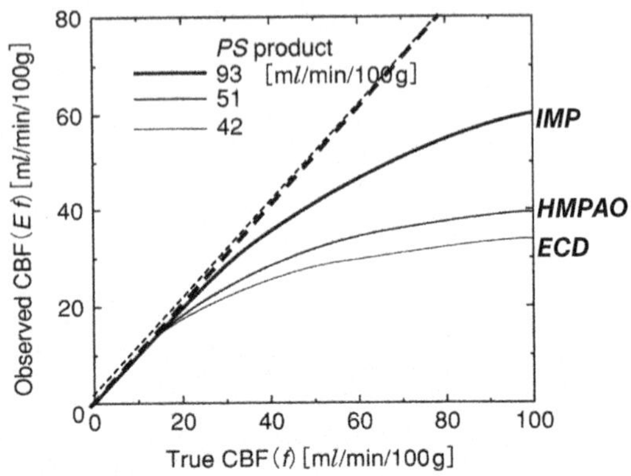

Fig. 5.11 Underestimation of cerebral blood flow measured with chemical microsphere tracers: Since the first pass extraction of an actual chemical microsphere tracer is $E < 1$, the cerebral blood flow measured by these tracers is underestimated. (Modified from Iida H et al. J Cereb Blood Flow Metab 1996)

5.3.3 Calculating Cerebral Blood Flow Using a Transit-Type Tracer

For a transit-type tracer, cerebral blood flow is calculated from the mean transit time (MTT). According to the aquarium model, after administration the tracer concentration immediately reaches equilibrium and the tracer then dilutes in accordance with the inflow rate of fresh blood (Fig. 5.3). The clearance (i.e., dilution) rate of the tracer concentration is proportional to the inflow rate *F* (mL/min) and inversely proportional to the aquarium volume *V* (mL). In other words, the clearance rate (or decay constant) is $k = F/V$ (min^{-1}). Since the effective-volume of the diffusible tracer is multiplied by the partition coefficient, the obtained clearance rate can be converted to cerebral blood flow by multiplying it by the partition coefficient. Consequently,

$$\text{Cerebral blood flow}(\mathrm{mL/g/min}) = \text{Clearance rate}\,k(\mathrm{min}^{-1}) \times \text{Partition coefficient}\,p(\mathrm{mL/g})$$
$$= pk(\mathrm{ml/min}\,E/\mathrm{g})$$

The mean transit time (MTT) is the inverse of the clearance rate (Fig. 5.5). Note that we usually think of brain tissue in 100 g units instead of 1 g. This is because it is convenient for clinical applications to express the cerebral blood flow as a full integer instead of using decimal places. This means that the cerebral blood flow for an average healthy man can be written as 50 mL/min/100 g rather than 0.5 mL/min/g. The convention of expressing cerebral blood flow in units of mL/min/100 g has been commonplace since Kety's era.

Like diffusible tracers, cerebral blood flow is calculated for nondiffusible tracers using the reciprocal of the MTT and the vascular volume (Fig. 5.5). However, since the mean transit time (MTT) of a nondiffusible tracer through the brain is only 10 s at most, which is of the same order of magnitude as the transit time of blood, it is critical to measure the MTT accurately. In cerebrovascular diseases, cerebral blood flow is greatly affected by collateral blood vessels, so we need to pay great attention to the vascular pathway. For example, in the case of unilateral cerebrovascular deficiency in a main trunk artery, cerebral blood flow in the diseased side may be affected by collateral circulation. In such a situation the true MTT cannot be measured, and the cerebral blood flow might be misestimated on either the left or right side of the brain.

5.3.4 Calculating Cerebral Blood Flow Using a Plug-Type Tracer

For a plug-type tracer, cerebral blood flow is calculated based on a different principle from the method for transit-type tracers. Since the plug-type tracer plugs the peripheral capillary region in proportion to the blood flow, cerebral blood flow is calculated from the density of tracer plugging the tissue (Rudolph and Heymann 1967). By measuring the tracer density, cerebral blood flow can be calculated (Fig. 5.12). Since the microspheres remain plugged in the capillaries, the density of microsphere plugs can be steadily measured with an excellent signal-to-noise ratio. For the traditional microsphere tracers that are only used in animal experiments, it is usually assumed that the first-pass extraction of the microsphere tracer is $E = 1$.

In the case of chemical microsphere tracers that are used clinically, the first-pass extraction E is less than 1, and E varies with CBF as described above (see Sect. 5.3.2). Since the actual enzyme activity and PS product for different chemical microsphere tracers vary according to the clinical procedure and SPECT scanner, it is necessary to

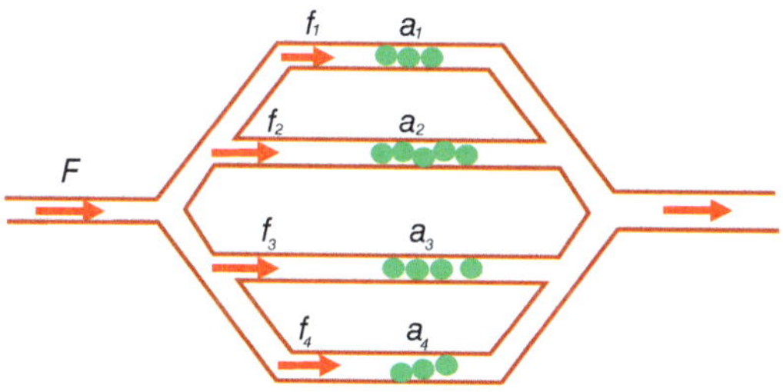

Fig. 5.12 Calculation of cerebral blood flow using a microsphere tracer: A microsphere tracer branches proportional to the blood flow at each bifurcation, so the cerebral blood flow can be calculated based on the number of microspheres plugging the vessels

setup a database of the expected tracer distribution for healthy subjects so that it can be used as a reference during diagnosis. It turns out that each tracer has its own specific regional cerebral blood flow distribution (Fig. 5.13) (Ito et al. 2006). Such a standard reference for each tracer in healthy subjects is needed as a control when used for clinical diagnosis based on cerebral blood flow patterns.

5.3.5 Absolute Measurement Versus Relative Measurement

Cerebral blood flow is a physiological index for the state of a subject or patient. For example, neurosurgeons expect an absolute value for the cerebral blood flow (i.e., in units of mL/100 g/min) to determine treatment strategies and evaluate the effect of the treatment. Absolute cerebral blood flow means that the measurement is not affected by factors such as the measuring device, the subject, or the measuring conditions and that the same result is obtained every time the measurement is made. That is, the reproducibility of the measurement is guaranteed. Throughout the history of cerebral blood flow measurement, the goal has always been to develop methods that provide absolute, or quantitative, results. To exclude factors that deteriorate the quantitative ability of the measurement technique, a small amount of invasiveness (e.g., direct arterial sampling to obtain the accurate input function) has been tolerated. However, in clinical stage measuring cerebral blood flow it is desirable to have a noninvasive method that places as little burden on the patient as possible while still producing the desired level for quantification.

Unfortunately, absolute quantification is not always possible so a relative, or qualitative, method is often used as the next best option. Relative methods include using a specific brain region as a reference region and evaluating the cerebral blood flow in the rest of the brain relative to the reference. For example, if cerebrovascular disease occurs in only one hemisphere of the brain, the healthy hemisphere is used as the reference region and the ratio of the diseased hemisphere to the healthy hemisphere is

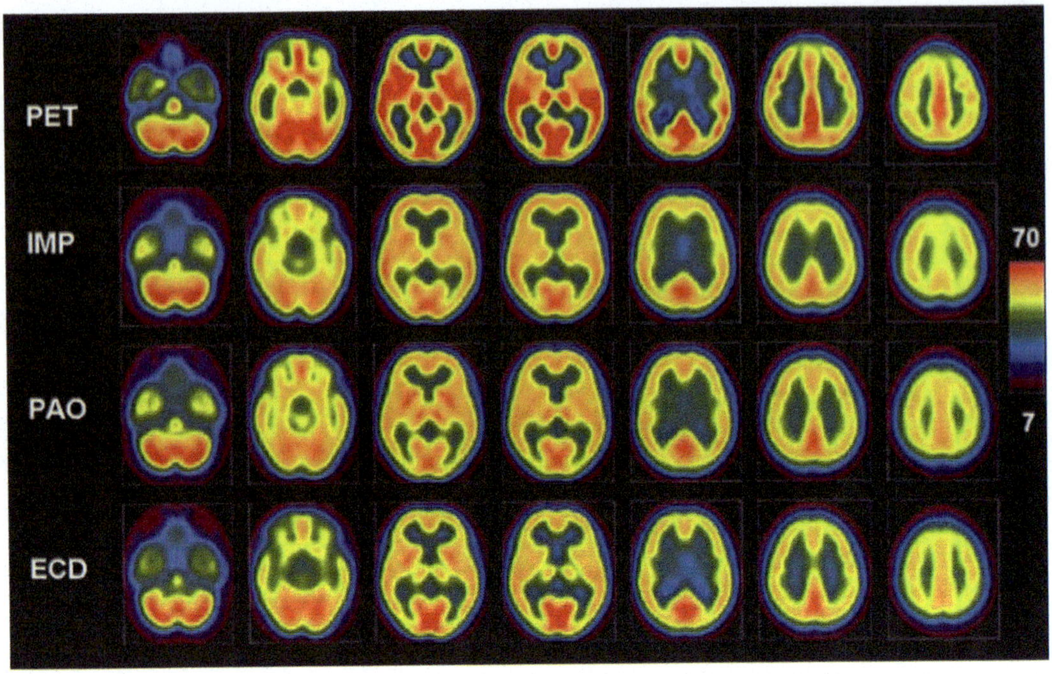

Fig. 5.13 Cerebral blood flow database for various SPECT tracers: Since each tracer has its own characteristics, the database must be used when diagnosing brain diseases. Cerebral blood flow obtained with PET and oxygen-15-labeled water is shown on the top row as the reference. (Ito H. et al. Ann Nucl Med 2006)

used as a simplified evaluation method. Methods of this type are called relative or qualitative measurements.

Thus, even though absolute/quantitative measurements are always desirable, sometimes only relative/qualitative measurements are possible. In the case of PET measurement of cerebral blood flow with oxygen-15-labeled water, the following is an example comparing absolute analysis and relative analysis (Fig. 5.14). The left column shows the cerebral blood flow measured during normal breathing (N; normoxia), the middle column shows the cerebral blood flow during hyperventilation (H; hyperventilation), and the right column is the subtraction of the left column from the middle column (H-N) superimposed on a CT image. For the absolute analysis (upper row), the subtraction image is negative across the whole brain due to hypocapnia during hyperventilation. However, for the relative analysis (lower row), where the whole-brain average has been normalized to be 50 ml/100 g/min, the motor cortex corresponding to the diaphragm is positive on the subtraction image in the right column. In this way, the activation foci in the motor cortex can be seen only in the relative analysis.

In nuclear medicine, absolute/quantitative or relative/qualitative analysis is performed according to the purpose of the examination. Of course, the absolute value of the cerebral blood flow is desirable for diagnosis. Nevertheless, in SPECT if it is difficult to obtain absolute values due to, for example, a lack of arterial blood sampling, diagnosis is often based on relative values (see Sect. 6.1.3). Many software applications have been developed to diagnose dementia based on the relative tracer distribution pattern of the cerebral blood flow.

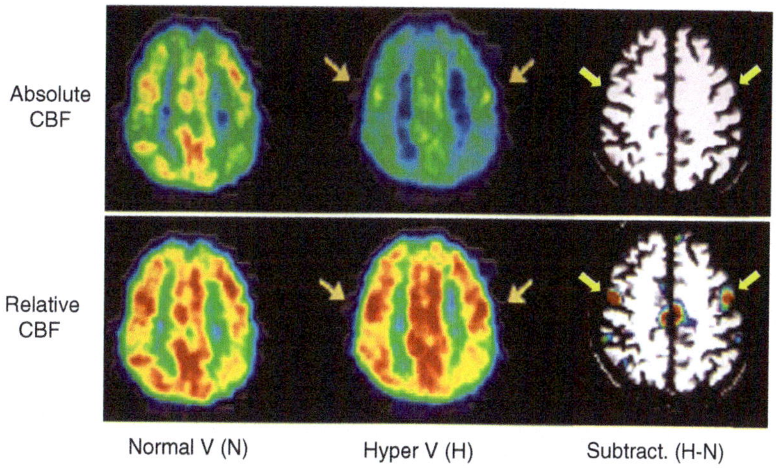

Fig. 5.14 Absolute analysis vs relative analysis: Cerebral blood flow (CBF) was measured during normal breathing (N) (left) and hyperventilation (H) (middle), and the subtraction of the two CBF images is superimposed on a CT image (H-N) (right). For the absolute CBF (upper row), as CBF during hyperventilation is low (middle), there is no foci appearing in the subtraction (right), while for the relative CBF (lower row), where all the images are normalized to the global average, activation foci (arrow) appear in the diaphragm motor area on the subtraction image

5.3.6 Semiquantitative Evaluation of Tracer Distribution

In PET, FDG accumulation is often evaluated semiquantitatively in examinations for the detection of malignant tumors. Without arterial blood sampling the FDG accumulation cannot be turned into an absolute measurement of glucose metabolism. The Standard Uptake Value (SUV), which is the FDG accumulation normalized by the FDG dose per body weight, is used instead. The more malignant the tumor, the higher the SUV. So, SUV is a good indicator of tumor grade.

Tracer accumulation can also be evaluated using the standard uptake value ratio (SUVR). The SUVR is calculated after selecting a reference region, and then taking the ratio of the tracer accumulation in a ROI to the tracer accumulation in the reference region. The SUVR is used when the administration dose or body weight are unavailable. When determining the density and affinity of neuroreceptors using neuroreceptor ligands, the SUVR is used to choose a reference region, such as the cerebellum, where there is no specific binding. The SUVR time series calculated from dynamic measurements can be used to evaluate the dynamic behavior of the tracer.

5.4 Cornerstone of Nuclear Medicine: RI Tracers

RI tracers are the cornerstone of nuclear medicine. There are many RI tracers in addition to those used for cerebral blood flow measurements. The general attributes required for RI tracers are summarized here.

5.4.1 Tracer Amount

An important prerequisite when using a tracer to measure a biological system is that the administered "tracer amount" has no physiological or biochemical effect on the measured system. In other words, it is assumed that the tracer will not disturb the biological system to be measured. For example, if you are measuring the temperature of a liquid in a cup with a thermometer, you measure the temperature by inserting the thermometer into the cup, but if the thermometer is large, inserting the thermometer may change the temperature of the liquid in the cup. If the thermometer disturbed the system, the measurement does not meet the requirements for a "tracer amount".

As another example, in arterial angiography a contrast agent is rapidly injected (1–5 s) in a dose approximately equivalent to a few percent of the carotid blood flow. If the contrast agent affects the blood flow in the carotid artery in some way, which in turn may affect the cerebral blood flow, the "tracer amount" condition is not valid. On the other hand, when measuring cerebral blood flow using laughing gas (N_2O), the laughing gas has negligible physical effect on cerebral blood flow so it can be said that the "tracer amount" condition is satisfied.

5.4.2 Specific Radioactivity

Specific radioactivity is an important property of RI tracers. In general, the specific radioactivity (Bq/mol) is defined as the amount of radioactivity (Bq) relative to the amount of the RI tracer compound (mol). For example, when a RI tracer is used to measure the distribution of low-density neuroreceptors in the brain, a high specific radioactivity is necessary. The more nonradioactive compound there is in the RI tracer, the less likely it is that the RI will bind to the target to be measured, which may cause the neuroreceptor density to be incorrectly estimated.

It is essential to reduce the amount of nonradioactive (or cold) compounds in the RI tracer so that it does not affect the model analysis. Thus, it is important to increase the specific radioactivity of the RI tracer. The specific radioactivity of the RI tracers produced by the PET facility at NIRS, where the author worked, is one of the highest in the world, reaching a maximum of 7.4 TBq/μmol. This is more than two orders of magnitude higher than the specific radioactivity produced by ordinary PET facilities, making the tracers from NIRS a powerful tool for detecting the density of neuroreceptors in ultrasmall amounts.

5.4.3 "Tracer" Versus "Probe"

Thus far, we have referred to RI-labeled drugs as "tracers", implying that they track the target. The goal is to measure biological functions by tracking the change in concentration and behavior of the tracer. However, some recent PET imaging research has focused on searching for the presence or absence of specific cells or molecules in the body. Therefore, the synthesis of RI-labeled tracers is increasingly designed to search for specific neurotransmitters, neuroreceptors and transporters, as well as specific amino acids and proteins. As such, the role of RI-labeled drugs in searching for cells and molecules rather than tracking is increasingly important, and in this context they are often called "probes" in the sense of exploration. Therefore, a RI-labeled drug may be called either a "tracer" or a "probe" depending on the purpose of a PET study. In this book, we will continue to use "tracer" unless otherwise noted.

5.4.4 Is the Detection Sensitivity of Nuclear Medicine Higher Than Other Modalities?

The sensitivity of the various measurement modalities can be compared. PET and SPECT nuclear medicine are said to have the highest sensitivity because the amount of tracer that can be detected is the smallest compared to other modalities. In nuclear medicine with PET and SPECT, the detection limit is said to be 10^{-9}, which means that the target volume in the field-of-view can be measured if there is a RI-labeled tracer at a density of 10^{-9}. In contrast, MRI requires a density of 10^{-5} in the target volume. This means that RI tracers can be used in concentrations four orders of magnitude less than that required for an MRI tracer. To say it in another way, PET and SPECT are four orders of magnitude more sensitive than MRI. CT requires a much greater amount of contrast agent than MRI, so the measurement sensitivity is even lower.

5.5 Compartment Models for Handling the Tracer Behavior

When using RI tracers in PET or SPECT, a compartment is a mathematical virtual room that corresponds to a physiological or biochemical unit. It is not visible, but it is an important virtual concept. In general, RI tracer behavior is analyzed by dividing the target region into multiple compartments according to its biochemical properties. The most important assumption when using a compartment model is that the physiological and biochemical parameters of the system (e.g. cerebral blood flow, metabolic rate, binding rate, dissociation rate, etc) must remain steady and invariable throughout the measurement.

5.5.1 Compartment: A Virtual Room Where the Tracer Is Distributed

Compartments are defined as virtual rooms that are separated biochemically or physically. The RI tracer moves between the compartments, and the speed at which it does so is called the rate constant. The rate constant is determined by physical driving factors, such as concentration differences between the compartments, and biochemical driving factors, such as affinity, metabolism, and dissociation (Fig. 5.15). Since the RI tracer is supplied through arterial blood, the tracer concentration in arterial blood is called the input function. Multiple compartments are set up in the brain depending on the nature of the tracer, but the differential equations describing the system become difficult to solve when there are more than three compartments in the brain. The model parameters, such as the rate constants, are esti-

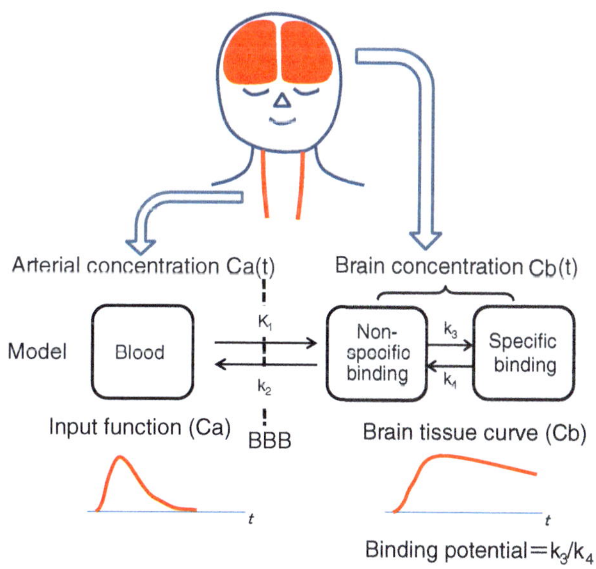

Fig. 5.15 Compartment model for neuroreceptor measurements: The figure shows a model with one blood compartment and two tissue compartments. The tracer concentration of the blood compartment, *Ca*, is measured with arterial blood sampling and is used as the input function to the brain. The tracer concentration of brain tissue, *Cb*, is the sum of the tracer concentration in the non-specific binding compartment, *Cn*, and the tracer concentration in the specific binding compartment, *Cs*, which are measured together in a tomogram ($Cb = Cn + Cs$)

mated from the solutions of the differential equations using appropriate boundary conditions, input functions measured from arterial blood samples, and PET brain tissue measurements of tracer concentration. The aquarium model (Fig. 5.3) is the simplest compartment model that can be used to calculate cerebral blood flow, as it has only one compartment in the brain.

5.5.2 Compartment Models for the Oxygen-15 Gas Inhalation Method

This compartment model is used to measure cerebral blood flow during the continuous inhalation of oxygen-15-labeled gas, which was the first practical application of PET. To measure cerebral blood flow, oxygen-15-labeled CO_2 is inhaled continuously. The oxygen-15-labeled CO_2 is instantaneously converted to oxygen-15-labeled water by carbonic anhydrase in the lungs and thereafter it circulates in the body as a diffusible tracer (West et al. 1961). If the concentration of radioactivity in arterial blood at time t is Ca, the amount of radioactivity flowing into the brain tissue is the product of Ca and the cerebral blood flow f, i.e., $f\ Ca$. On the other hand, the decrease of radioactivity in the brain tissue is the sum of two components: outflow and physical decay. If the brain tissue concentration at time t is Ci, the outflow component is Ci multiplied by the cerebral blood flow and divided by the partition coefficient, i.e. $(f/p)\ Ci$, while the second component is the decrease in radioactivity due to the physical decay of oxygen-15-labeled O_2, i.e. $\lambda\ Ci$. Here, λ is the physical decay constant of oxygen-15. When enough time has elapsed (about three times the half-life of oxygen-15, or approximately 10 min), the inflow, outflow and decay balance out so that the time variation (time differentiation) of Ci is zero (Fig. 5.16). That is,

$$\mathrm{dCi}/\mathrm{dt} = f\,Ca - (f/p)Ci - \lambda\,Ci = 0$$

If the blood radioactivity concentration Ca is known from arterial blood sampling and the brain tissue radioactivity concentration Ci is known from PET measurements, the cerebral blood flow f can be estimated with

$$f = \lambda\,\mathrm{Ci}/(Ca - \mathrm{Ci}/\mathrm{p})$$

If this result is combined with PET measurements made using continuous inhalation of oxygen-15-labeled O_2, then cerebral oxygen metabolism can also be estimated. However, it is necessary to use an extended compartment model for that purpose (Fig. 5.17). The oxygen-15-labeled O_2 binds to hemoglobin in the alveoli and is transported throughout the body. Only a fraction of the oxy-

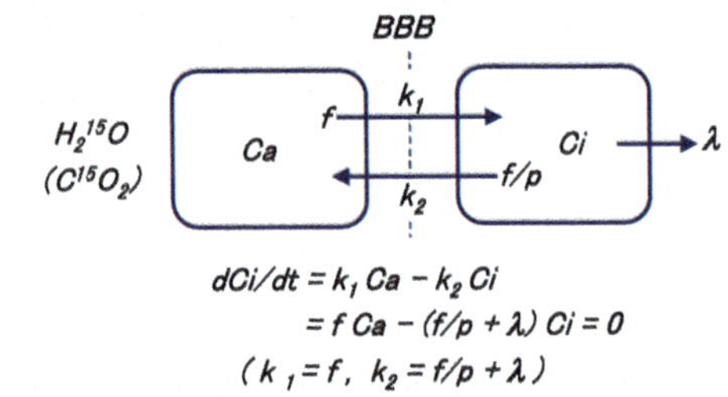

Fig. 5.16 Steady-state inhalation model for ^{15}O-labeled CO_2: The concentration of radioactivity (Ci) accumulated in brain tissue depends on the difference between inflow and outflow. The inflow corresponds to $H_2{}^{15}O$ ($f\,Ca$) entering from arterial blood. The outflow consists of $H_2{}^{15}O$ ($f/p \cdot Ci$) returning to the blood, and the physical decay of $H_2{}^{15}O$ ($\lambda \cdot Ci$). Since these are balanced in the steady-state condition, $f\,Ca = (f/p + \lambda)\,Ci$. Cerebral blood flow ($f$) is obtained after solving this equation

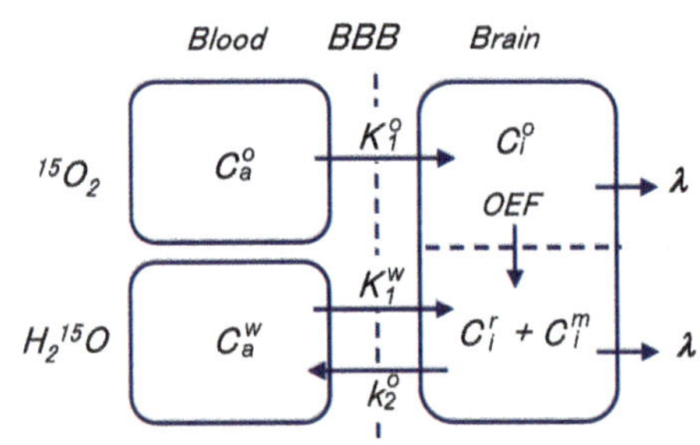

Fig. 5.17 Steady-state inhalation model for ^{15}O-labeled O_2: When ^{15}O-labeled O_2 gas is inhaled, the extracted oxygen fraction (OEF) of ^{15}O-labeled O_2 is metabolized to ^{15}O-labeled metabolic water (Cmi). In addition to this, there is a contribution from ^{15}O-labeled water (Cri) recirculating throughout the body. The sum of the ^{15}O-labeled water ($Cri + Cmi$) also undergoes physical decay. The OEF can be estimated from the balance of these contributions

gen, called the oxygen extraction fraction (OEF), is metabolized in the brain to produce oxygen-15-labeled water. The radioactivity from oxygen-15-labeled water in the brain reaches a balance between the metabolized water in the brain, the metabolized water flowing into the brain from the body, the metabolized water flowing out of the brain due to cerebral blood flow, and physical decay. Based on this balance, PET measurements of the oxygen-15-labeled water concentration in the brain at steady-state, the arterial blood radioactivity concentration from blood sampling and the measured cerebral blood flow, the oxygen extraction fraction (OEF) and cerebral oxygen metabolism ($CMRO_2$) can be estimated with the extended compartment model.

5.5.3 Compartment Model for Neural Transmission

Here we consider a compartment model for tracers that is often used in PET analysis to measure the density of neural receptors, transmitters and transporters. There are multiple compartments in the brain, but the simplest model assumes two compartments: one for nonspecific binding of the tracer to receptors other than the target receptor and the other for specific binding to the target receptor (Fig. 5.15). The binding potential, which indicates the receptor density, is calculated from the ratio of the rate constants describing movement of the tracer between the two compartments. The arterial input function, which will be discussed in the next section, is essential for model analysis, but direct measurement of the input function requires arterial blood sampling, which is invasive, so we devise a substitute. In general, the rate constants for nonspecific binding are much faster than those for specific binding, so the tracer concentration in the nonspecific binding compartment is sometimes used instead of the arterial input function when analyzing the differential equation. To do that it is necessary to identify a region of the brain where no binding occurs. For example, as it is well known that the cerebellum has no binding site for dopamine, the cerebellum is usually used as the reference region in analysis for dopamine ligand. If a new RI ligand is developed, the kinetics is analyzed pixel-by-pixel for the whole brain and the largest group of adjacent pixels with similar rate constants will be selected as the reference region.

5.5.4 Arterial Input Function of the Tracer

Since the RI tracer behavior is analyzed with differential equations, the arterial blood input function of the tracer supply to the brain must be accurately measured. The most accurate way to do this is to invasively draw arterial blood and directly measure the radioactivity concentration in that blood. However, such an invasive method is now considered unethical unless there is an unavoidable need. Thus, alternative methods are usually devised. The following describes some of the noninvasive methods used to estimate the arterial input function.

5.5.4.1 Indirect Measurement of the Input Function

As an alternative to direct blood sampling, one noninvasive way to estimate the input function is to measure a physiologically equivalent index of the arterial input function. For example, in the case of inert noble gases that enter and exit through the lungs, such as xenon-133, the exhaled gas concentration is in equilibrium with the concentration in arterial blood. Therefore, the exhaled gas can be collected and the end-expiratory radioactivity dynamically measured with a separately prepared detector to estimate the arterial radioactivity. A noninvasive method to determine the input function is also an issue for PET measurements.

5.5.4.2 Estimating Arterial Blood Radioactivity Based on PET Images

An image-based method to estimate the input function has been developed for PET. This is known as the image-guided estimation of carotid artery input function method. Recent

PET systems have a long axial field-of-view of more than 20 cm so that the whole brain and carotid artery can be measured simultaneously. The radioactivity in the carotid artery can therefore be measured dynamically, and the input function can be estimated from these measurements. This method requires additional high spatial resolution morphology data, usually MRI or CT images, to determine the position of the carotid artery. Unfortunately, the function may be underestimated due to the partial volume effect and signal spilling over from the neighboring jugular veins (spillover effects). That is why it is necessary to verify that we are measuring a pure carotid input function. In the past, the author and colleagues developed a dual PET system where two PET scanners were arranged on the same axis. One scanner was positioned over the brain and the other was placed over the upper trunk so that the radioactivity in the heart cavity could be directly measured (Iida et al. 1998).

5.5.4.3 Estimation from a Reference Region in Brain Tissue

As mentioned earlier, instead of the arterial input function, the behavior of PET tracers designed to measure neuroreceptor density can be analyzed using a brain region where there is nonspecific binding. For example, in the case of dopamine receptor tracers, it is known that there are few dopamine receptors in the cerebellum, so a reference region analysis method has been developed that uses the signal from the cerebellum as the input function. On the other hand, some tracers do not have obvious brain regions where there is nonspecific binding so a reference region cannot be easily set. Additionally, the reference region approach cannot be used when developing a new tracer because it is unknown where, or even whether, there is a nonspecific binding region that is suitable as a reference. In this case, invasive arterial blood sampling may initially be necessary to identify a suitable nonspecific reference region.

References

Herscovitch P, Raichle ME. What is the correct value for the brain—blood partition coefficient for water? J Cereb Blood Flow Metab. 1985;5(1):65–9.

Hoedt-Rasmussen K, Sveinsdottir E, Lassen NA. Regional cerebral blood flow in man determined by intra-arterial injection of radioactive inert gas. Circ Res. 1966;18(3):237–47.

Iida H, Akutsu T, et al. A multicenter validation of regional cerebral blood flow quantitation using [123I]iodoamphetamine and single photon emission computed tomography. J Cereb Blood Flow Metab. 1996;16(5):781–93.

Iida H, Miura S, et al. Noninvasive quantitation of cerebral blood flow using oxygen-15-water and a dual-PET system. J Nucl Med. 1998;39(10):1789–98.

Ito H, Inoue K, et al. Database of normal human cerebral blood flow measured by SPECT: I. Comparison between I-123-IMP, Tc-99m-HMPAO, and Tc-99m-ECD as referred with O-15 labeled water PET and voxel-based morphometry. Ann Nucl Med. 2006;20(2):131–8.

Kanno I, Lassen NA. Two methods for calculating regional cerebral blood flow from emission computed tomography of inert gas concentrations. J Comput Assist Tomogr. 1979;3(1):71–6.

Kanno I, Masamoto K. Bridging macroscopic and microscopic methods for the measurements of cerebral blood flow: Toward finding the determinants in maintaining the CBF homeostasis. Prog Brain Res. 2016;225:77–97.

Phelps ME, Huang SC, et al. Cerebral extraction of N-13 ammonia: its dependence on cerebral blood flow and capillary permeability—surface area product. Stroke. 1981;12(5):607–19.

Renkin EM. Capillary and cellular permeability to some compounds related to antipyrine. Am J Phys. 1953;173(1):125–30.

Rudolph AM, Heymann MA. The circulation of the fetus in utero. Methods for studying distribution of blood flow, cardiac output and organ blood flow. Circ Res. 1967;21(2):163–84.

West JB, Dollery CT, et al. The use of radioactive carbon dioxide to measure regional blood flow in the lungs of patients with pulmonary disease. J Clin Invest. 1961;40:1–12.

Measuring Cerebral Blood Flow: Methods

6

Abstract

This chapter describes practical methods to measure cerebral blood flow. For SPECT, the diffusible tracers Xe-133 and I-123 IMP are described. These are used for quantitative measurements of cerebral blood flow to evaluate the necessity for treatment in stroke. Chemical microsphere tracers labeled with Tc-99m are widely used for semiquantitative cerebral blood flow measurements. In PET, quantitative cerebral blood flow and metabolism can be measured using O-15-labeled gases and FDG. Neural transmission and abnormal protein deposits are examined for degenerative diseases using some newly developed tracers. This chapter also discusses methods that do not use RI tracers. These include cold xenon as a diffusible tracer with CT, arterial spin-labeling with MRI, and contrast agents as nondiffusible tracers with both CT and MRI. In addition, several experimental methods for animals using laser Doppler flowmetry (LDF), laser speckle imaging (LSI) and two-photon microscopy are introduced.

Keywords

SPECT · Xe-133 · I-123 · Tc-99m · PET · O-15 · Arterial spin-labeling · Near infrared spectroscopy · H_2 clearance · Autoradiography · Compact cyclotron

6.1 Measurement of Cerebral Blood Flow Using SPECT

This section summarizes methods for measuring regional cerebral blood flow using SPECT tracers. Over time, it became unethical to perform the cerebral clearance method using an invasive injection of xenon-133 dissolved in saline into the internal carotid artery. The xenon-133 gas inhalation method was therefore developed as an alternative. On the other hand, chemical microsphere tracers labeled with technetium-99m have been developed and, although they only measure semiquantitative cerebral blood flow, their clinical application has rapidly expanded due to the improved resolution of SPECT systems as well as the excellent signal-to-noise ratio.

6.1.1 Quantitative Cerebral Blood Flow Using Xenon-133

Up until the 1980s, the Lassen-Ingvar method of intracarotid artery injection of xenon-133 dissolved saline was used in most institutions because it was the only method that could measure quantitative cerebral blood flow (Lassen and Ingvar 1961). However, the method required an invasive puncture of the internal carotid artery, so to avoid this, Obrist developed the xenon-133 gas inhalation method (Obrist et al. 1967). Another advantage of the

I. Kanno, *What Cerebral Blood Flow Tells Us*, https://doi.org/10.1007/978-981-97-6382-5_6

xenon-133 inhalation method is that it can measure the cerebral blood flow of the entire brain, both cerebral hemispheres and the cerebellum, whereas the xenon-133 intracarotid arterial injection method can only measure either the left or right hemisphere of the cerebrum (Fig. 6.1). The disadvantage is that the arterial input function to the brain is temporally broadened. In the xenon-133 intracarotid arterial injection method, the tracer is administered in 1–2 s, so the arterial input function can be approximated as a unit impulse (mathematically defined as a pulse of zero width in time and infinite amplitude, but which has an integral value of unity). In that case the differential equations are unnecessary, and the brain clearance curve becomes a simple exponential function. In contrast, the xenon-133 gas inhalation method requires deconvolution correction of the brain clearance curve because the xenon-133 supply to the brain (after passing through the heart and pulmonary circulation), will occur over a period of 10–30 s. Obrist et al. used a least-squares approximation for deconvolution of the input function (Obrist et al. 1967). The input function is estimated from measurements of the xenon-133 concentration in end-tidal expired gas or from measurements of the radioactivity in the lung, but both methods require calibration to convert to an arterial concentration.

However, with the advent of SPECT, it became difficult to perform pixel-by-pixel deconvolution analysis due to the large amount of noise in each pixel of the image. A xenon-133 tomography system does not obtain sufficient count rates, so the signal-to-noise ratio of the tomogram is quite low. In addition, it takes several tens of seconds to obtain a single tomogram, so the sampling interval is sparse. As a result, direct analysis of the dynamic tracer concentration in each pixel is difficult. To solve this low signal-to-noise problem, two methods were proposed: (1) the sequential picture method, and (2) the early picture method (Kanno and Lassen 1979). When Lassen's group in Copenhagen, Denmark, was developing a xenon-133 brain SPECT with the goal of dynamic data collection using a rotating rectangular detector (see Sect. 4.3.4), they needed a method to solve the low signal-to-noise problem. The methods were also applied for the PET system of Yamamoto et al. at MNI, Montreal, which was using the diffusible positron-emitting noble gas krypton-77 (Yamamoto et al. 1977).

The sequential picture method assumed that the 1-min integrated images were measured in 4 ~ 5 consecutive frames and, assuming that the arterial input function was correctly measured, convolution integral curves were simulated for the range of possible cerebral blood flow levels. The area under each simulated curve was standardized before the standardized curves were

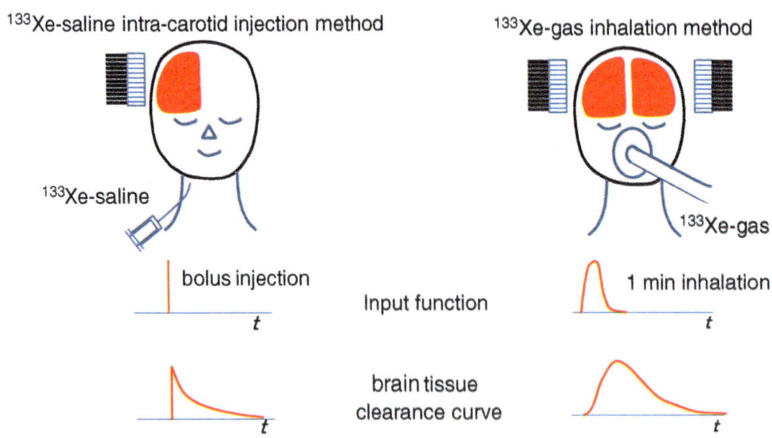

Fig. 6.1 Schemes for the xenon-133 clearance methods: (Left) The xenon-133 saline intra-carotid artery injection method measures cerebral blood flow in the hemisphere on the injected side. (Right) On the other hand, the xenon-133 gas inhalation method allows the cerebral blood flow of both hemispheres to be measured

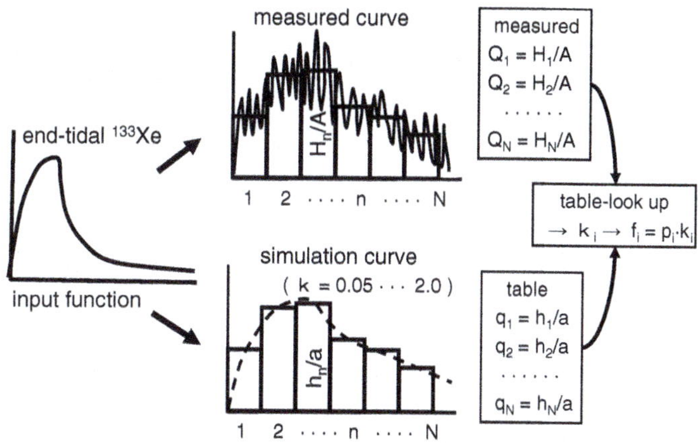

Fig. 6.2 The Kanno-Lassen sequential-picture method: A lookup table is prepared in advance for each cerebral blood flow level, and the cerebral blood flow is determined pixel-by-pixel using the table. See text for more detail. (Modified from Kanno, Lassen, J Comput Assist Tomogr 1979)

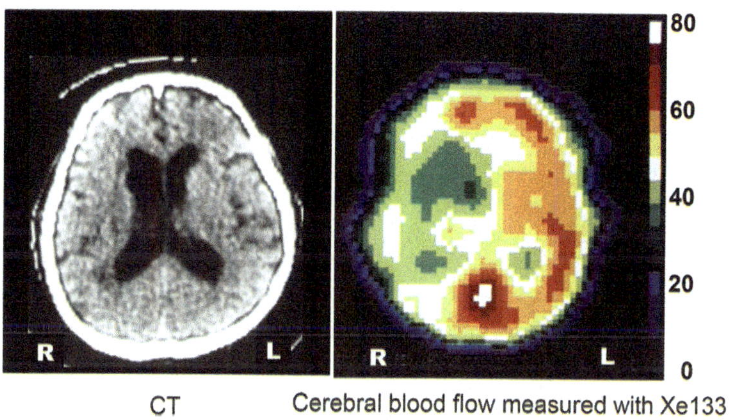

Fig. 6.3 The first tomographic cerebral blood flow image taken in Japan: Xenon-133 cerebral clearance data was analyzed with the Kanno-Lassen sequential-picture method for a patient with middle cerebral artery occlusion of two-months after the onset. A widespread deficit of cerebral blood flow was found in the right hemisphere. Unit of cerebral blood flow is ml/100g/min

sampled at intervals matching the measurement times and stored in a lookup table. The tabulated curves were then compared to the measured data and the local cerebral blood flow was determined as that corresponding to the best match (Fig. 6.2). The first quantitative tomographic cerebral blood flow measurement for a patient with cerebral infarction in Japan was made with this method (Fig. 6.3). It revealed a marked decrease in cerebral blood flow across a wide area, even in the deep structure of the brain, and enabled a more accurate prognosis of the disease than a nontomographic planar measurement.

The early picture method assumed that the initial tracer uptake was approximately proportional to cerebral blood flow. The integral of the initial part (1–3 min) of the clearance curve was simulated for the range of possible cerebral blood flows, and a nomogram was made in advance of the measurement. Using this nomogram, the regional cerebral blood flow was obtained by matching with the integral of measured clearance curves from the tomographic images (Fig. 6.4). This method has been widely applied and is frequently used in PET and other fields to convert the raw images of RI concentration into functional images.

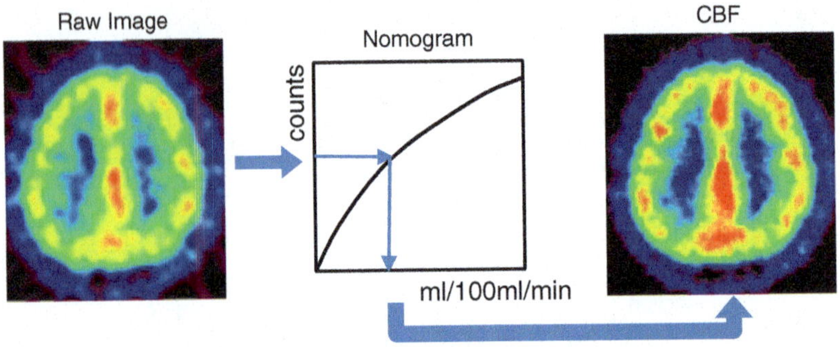

Fig. 6.4 Kanno-Lassen early-picture method: Assuming the initial tracer accumulation is proportional to cerebral blood flow, a nomograph is calculated for a range of possible cerebral blood flows. The integral of the raw radioactivity is converted pixel-by-pixel to the cerebral blood flow. See text for more detail

Column: Edda Sveinsdottir, a Lady Mathematician

The author developed the Kanno-Lassen method in Copenhagen during his first trip abroad in 1978. Most of this time was spent working on the development of the program with a lady mathematician, Edda Sveinsdottir. A few years earlier, in 1971, when the Congress for Brain Circulation was held in Tokyo, a delegation from the Nordic Brain Circulation Group visited the Akita Brain Institute. One of the delegates was a tall lady mathematician, and I remember feeling dazzled when I heard that she was Mrs. Lassen. In Copenhagen, I used the computer at the University of Copenhagen's Faculty of Science, where she worked, to solve a problem given to me by Lassen by performing a convolutional integral of the brain input function and brain tissue clearance curve to simulate measured data. I later learned that she and Lassen had divorced just before the author came to study in Copenhagen. However, they had coauthored many papers even after their divorce. They had kept their public and private lives separate to continue their collaboration. This lifestyle was a cultural shock for me. Such a thing was unthinkable in Japan in the 1970s. Edda kindly took care of my family, and we were able to experience the simple but luxurious life and culture of Scandinavian Denmark even though our stay of 6 months was relatively short.

6.1.2 Quantitative Cerebral Blood Flow Using Iodine-123-Labeled IMP

Iodine-123-labeled N-isopropyl-p-iodoamphetamine (IMP) was initially treated as a chemical microsphere tracer, but subsequent analysis revealed that it slowly returns to the bloodstream from brain tissue, indicating that it can be analyzed as a diffusible tracer with a slow diffusion rate, i.e., it has a very large partition coefficient (see Sect. 4.3.1). The slow diffusion rate means that it is very difficult to estimate the cerebral blood flow from the clearance curve. The autoradiograph method (ARG) was therefore developed. To calibrate the standard input function for this method a single arterial blood sample is required from the patient. The standard input function is a template made in advance as the average input function obtained from many subjects (Iida et al. 1994).

In addition, a method has been developed to quantitatively compare regional cerebral blood flow at rest and during acetazolamide (ACZ) loading (Kim et al. 2006). To aid these measurements, it is possible to split iodine-123-labeled IMP into two doses so that quantitative cerebral blood flow can be measured twice within a time of about 1 h. Even in the absence of PET, this split-dose method can be used to assess circulation reserve (see Sect. 7.3.2), thereby enabling treatment decisions to be made. From a physical point of view, iodine-123-labeled IMP enables a higher spatial resolution than xenon-133 due to its higher gamma-ray energy, and it can be measured for a longer time due to the slower clearance of tracer concentration.

Therefore, iodine-123-labeled IMP is widely used clinically as a tracer to measure quantitative cerebral blood flow in facilities without PET.

6.1.3 Qualitative Cerebral Blood Flow Using Chemical Microsphere Tracers

In the late 1990s, technetium-99m-labeled HMPAO and technetium-99m-labeled ECD became commercially available as chemical microsphere tracers for SPECT. With the introduction of these new tracers, clinical cerebral blood flow measurements quickly became popular. Compared to the xenon-133 brain clearance method, these new tracers provide cerebral blood flow imaging with higher spatial resolution and higher signal-to-noise ratio. For example, in the hyperacute phase of cerebrovascular disease after occlusion of the left internal carotid artery, a decrease in cerebral blood flow throughout the cerebral hemispheres can be observed at a glance (Fig. 6.5, right), and the difference is obvious compared to morphological CT images that do not show clear signs of cerebral infarction (Fig. 6.5, left). Even qualitative cerebral blood flow measurements clearly show brain abnormalities. The clinical impact of images like this was so significant that the technique quickly spread to major medical facilities. If the cerebrovascular deficit is confined in a unilateral hemisphere, the ratio of the cerebral blood flow in the left and right cerebral hemispheres can be used to estimate the degree of cerebral blood flow reduction in the diseased side, providing useful information for treatment planning in the acute stage of stroke. The left-right ratio is also useful in follow-up examinations of cerebrovascular diseases. Thus, when absolute quantitative cerebral blood flow cannot be obtained, substituting it with relative cerebral blood flow can be clinically useful. However, in the case of long-term follow-up examinations over several years, there are many cases where cerebral blood flow changes have occurred on a whole-brain basis due to aging etc. Therefore, care must be taken to avoid the pitfalls associated with using relative cerebral blood flow.

Cerebral blood flow is frequently used not only for the diagnosis of cerebrovascular diseases but also for the diagnosis of various degenerative diseases and dementias (see Sects. 7.4.1, 7.4.2, 7.4.3 and 7.4.4). Chemical microsphere tracer measurements in daily clinical practice provide relative cerebral blood flow, not absolute values, so most diagnosis software only needs relative values. It is necessary to prepare a database of normal subjects for each tracer at each

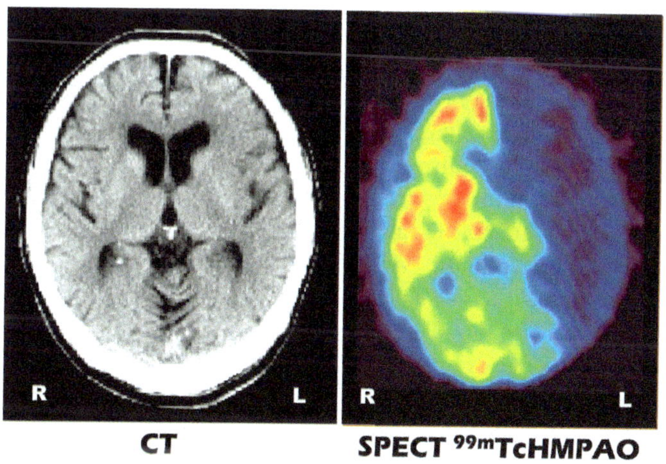

Fig. 6.5 Acute patient with an occlusion of the left internal carotid artery: (Left) There is no abnormality in the morphological CT. (Right) However, a clear and extensive decrease is shown in 99mTc-HMPAO cerebral blood flow in the left cerebral hemisphere

facility in advance because each tracer has its own characteristics (e.g., Tc-99m-labeled HMPAO and Tc-99m-labeled ECD have different first-pass extractions). Inputting an incorrect value can lead to underestimation at high cerebral blood flow or poor reproducibility at low cerebral blood flow. Therefore, it is necessary for each institution to prepare a database of normal subjects for each tracer in advance (see Fig. 5.13) (Ito et al. 2006).

6.2 Measurement of Cerebral Blood Flow and Metabolism with PET

PET uses positron-emitting radioisotopes (RIs). The measurement of cerebral blood flow is mainly performed using oxygen-15-labeled water. One of the features of oxygen-15-labeled water is that it has a short half-life of approximately 2 min, which enables the measurement procedure to be repeated in a short interval. In addition, PET measurements of cerebral blood flow and cerebral oxygen metabolism can be combined using oxygen-15-labeled gases.

6.2.1 The Emergence of the Compact Cyclotron

Although PET is reliable in terms of quantitative measurements, it requires a large investment in equipment and many staff to operate it. PET uses short half-life RIs: fluorine-18 (110 min), carbon-11 (20 min), nitrogen-13 (10 min), and oxygen-15 (2 min). A circular accelerator, or cyclotron, is required to generate RIs for a medical facility. Until the 1970s, cyclotrons were big machines that accelerated particles at high energy and required a dedicated building and many physics experts to operate the system. In the 1960s, there were only a few medical facilities with cyclotrons, including Hammersmith Hospital in London, Massachusetts General Hospital in Boston, and Washington University in St. Louis.

In the 1970s, RIKEN and Japan Steel Works (JSW) collaborated and succeeded in developing the world's first compact cyclotron (Fig. 6.6). Later, Sumitomo Heavy Industries in Japan and CGR MeV in France also jointly developed compact cyclotrons. In addition, a Swedish company and a venture company in the U.S. also began development. International competition led to the installation of compact cyclotrons in medical facilities around the world in the 1980s. Operation of a compact cyclotron can be easily automated with a computer. However, one of the two Japanese cyclotron manufacturers, JSW, the first to succeed in developing the compact cyclotron, was burdened with the large development costs and withdrew from the compact cyclotron market before it became profitable. Ironically, after the withdrawal of JSW it was widely recognized that FDG-PET was useful for the diagnosis of cancer, and health insur-

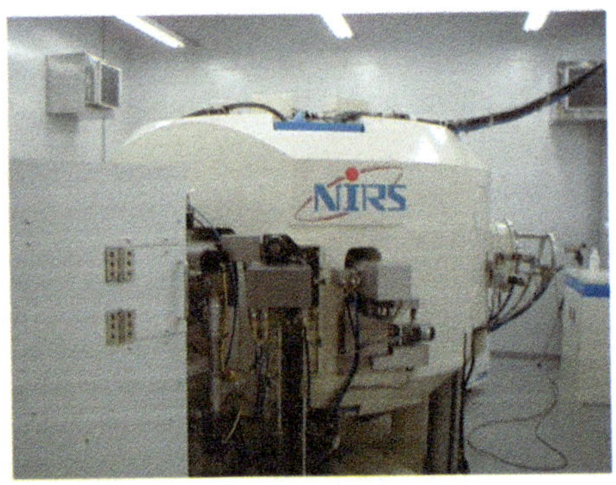

Fig. 6.6 The compact cyclotron (BC1810, Japan Steel Works Ltd) at NIRS: Protons and deuterons are accelerated, and the target is irradiated to produce positron-emitting radioisotopes of carbon-11, nitrogen-13, oxygen-15, and fluorine-18

ance coverage was introduced in 2002, opening the PET era. Today, two Japanese companies, Sumitomo Heavy Industries and JFE Engineering Corporation, which partially uses overseas technology, continue to supply the compact cyclotron market.

Column: Light and Dark in the Development of Compact Cyclotrons

A cyclotron is a circular accelerator. It used to be a large device requiring a whole building. However, since RIs for nuclear medicine can be produced with low acceleration energy, Karasawa and Nozaki, pioneering researchers at RIKEN, together with Japan Steel Works (JSW), started to develop a compact cyclotron that broke the conventional concept. It took several years to complete the prototype, and it was installed at Nakano National Hospital in 1976. Nozaki was a leader in chemistry, pharmacology, and medical research using cyclotrons. He was also the principal of the Cyclotron Summer School, which has been held every year since the late 70s and is now a large meeting. Today, compact cyclotrons are basic equipment for PET nuclear medicine, but in the early days we were still in the process of exploring what we could do with cyclotrons and how to use them. If JSW's decision to withdraw from the market had been made slightly later, the PET community in Japan would have been able to compete with foreign companies to innovate and modernize cyclotron technology, as well as to develop peripheral technologies for PET tracers. Most importantly, I believe that a variety of human resources would have been able to come together to create an industry of dreams.

6.2.2 Practical Application of PET Systems

When Ter-Pogossian visited Japan in the early 1980s to give a lecture on the potential of PET, I asked him about the expectations for PET. He immediately answered that all nuclear medicine examinations would be replaced by PET studies within the next 5 years. However, almost 40 years later his prediction has not yet come true. The number of PET facilities in Japan has increased since the health insurance system began supporting cancer screening with FDG, and more than 400 PET facilities are now in operation at major medical institutions. However, more than 90% of these facilities use FDG only for cancer screening, and only a few research facilities and university hospitals use PET for other purposes, such as cerebral blood flow. SPECT is still the predominant method in clinical practice, but it is becoming more common to use either SPECT or PET depending on the measurement goal. It is likely that PET and SPECT will continue to be used together, with each having its own positive characteristics. Although Ter-Pogossian's prediction was overly optimistic, he clearly saw a bright future for nuclear medicine.

Column: Ter-Pogossian, Pioneer of PET

Ter-Pogossian was a pioneer in PET development. He visited the author's institute several times and I had many opportunities to talk with him. The Mallinckrodt Laboratory at the University of Washington, to which he belonged, had a large medical cyclotron and was one of the leading institutions in the USA for nuclear medicine research. However, Ter-Pogossian came to Japan to investigate a compact cyclotron that could be easily used for PET. He visited the Akita Brain Institute where the JSW cyclotron was in operation. At that time, we had just started PET operations and he praised our PET study of cerebral circulation and metabolism using an oxygen-15-labeled tracer. On the other hand, Ter-Pogossian, like Lassen, also had a diamond head. He claimed that biochemically, deoxyglucose (DG) is toxic because it is a metabolic inhibitor in the brain, so it is impossible to measure the rate of glucose metabolism with DG. He also insisted that the accumulation of fluorine-18-labeled FDG in brain tumors was not due to metabolism but was merely the result of BBB breakdown within the tumor (see Sect. 3.2.1). Ter-Pogossian died in Paris in 1996 while on a trip, but I believe he maintained this claim to the end.

Fig. 6.7 Michel M Ter-Pogossian (1925–1996)

6.2.3 Steady-State Inhalation Method with Oxygen-15-Labeled Gas

Washington University in St. Louis, USA, and Hammersmith Hospital in the UK were the first to use the oxygen-15-labeled gas produced by a cyclotron for medical purposes. At Washington University, Ter-Pogossian (Fig. 6.7) and colleagues tried to measure both oxygen metabolism and cerebral blood flow by placing a detector on the head and recording the radioactivity concentration in the brain after inhaling oxygen-15-labeled O_2 (Fig. 6.8) (Ter-Pogossian et al. 1970). At Hammersmith Hospital in the UK, West and colleagues established the basic behavior of oxygen-15-labeled gas. West showed that when oxygen-15-labeled CO_2 is inhaled, it is instantaneously converted to oxygen-15 water by carbonic anhydrase in the lungs, and that oxygen-15-labeled water gradually increased in the blood (West et al. 1961).

After studying in the environment at Hammersmith Hospital, Terry Jones went to the United States and developed the principles for measuring cerebral blood flow and cerebral oxygen metabolism after steady-state inhalation of oxygen-15-labeled CO_2 and oxygen-15-labeled O_2, respectively (Jones et al. 1976). Later, he returned to Hammersmith Hospital and developed a quantitative method for PET measurement of cerebral blood flow, cerebral oxygen metabolism, and oxygen extraction fraction (Fig. 6.9) (Frackowiak et al. 1980). Furthermore, the addition of carbon-11-labeled CO inhalation made it possible to simultaneously estimate the cerebral blood volume (Lammertsma and Jones 1983). As mentioned earlier, the steady-state inhalation method measures the balance between the inflow of oxygen-15 and its physical decay (see Sect. 5.5.2). Including prep-

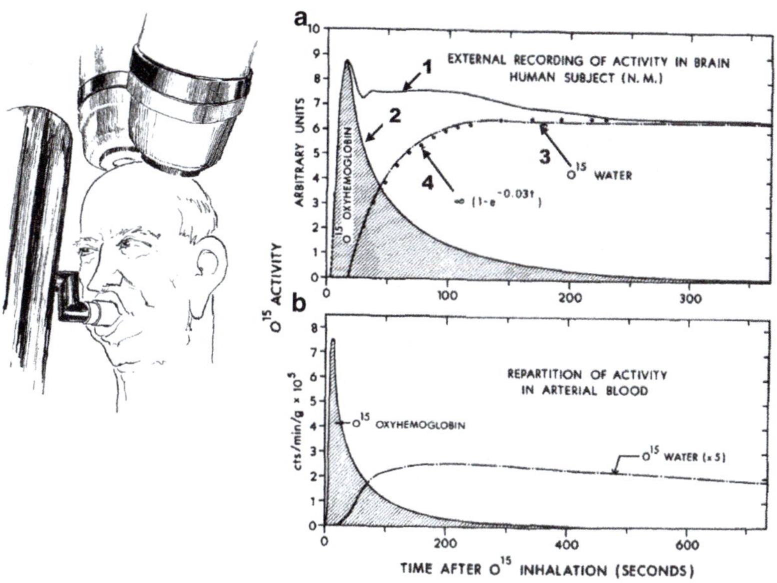

Fig. 6.8 Residual radioactivity after inhalation of oxygen-15-labeled O_2 gas: (**a**) Distribution of oxygen-15 radioactivity in the brain measured with the detector over the parietal area (1), the hemoglobin bound radioactivity measured from sampled carotid blood (2), and the difference between curve 1 and curve 2 (3), which represents water bound activity. (**b**) Radioactivity in the carotid blood corresponding to oxyhemoglobin and metabolized water. (Ter-Pogossian and Herscovitch, J Clin Invest, 1970)

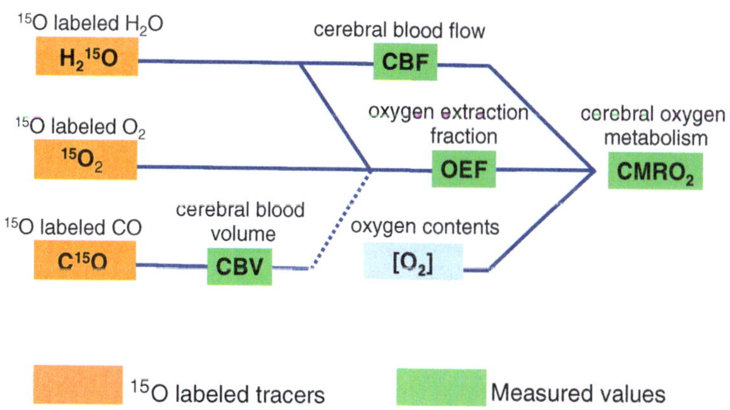

Fig. 6.9 Relationship between the parameters in the PET ^{15}O steady-state inhalation model: The orange boxes correspond to the administered tracers. The green boxes are the measured parameters. The gray box is the parameters measured from blood sampling

aration time, the measurements take more than 2 h to perform. In addition, the method was very invasive because arterial sampling was required. Nevertheless, since it was the only method at the time that could quantitatively measure cerebral blood flow, cerebral oxygen metabolism, and oxygen extraction fraction in the resting state (Fig. 6.10), it was widely applied around the world to investigate the pathogenesis of cerebrovascular disorders and various brain diseases.

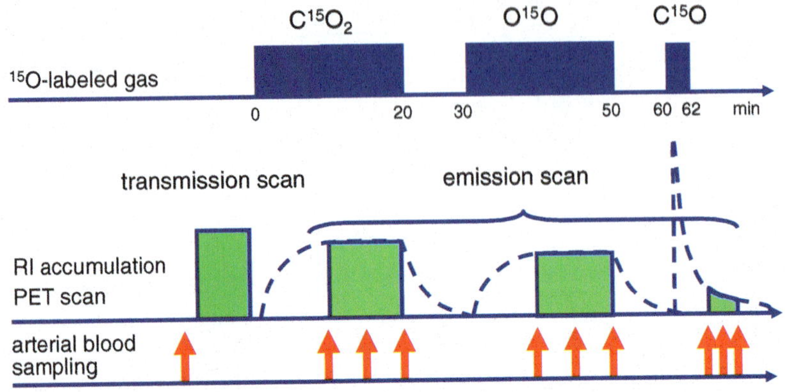

Fig. 6.10 Protocol for the oxygen-15-labeled gas steady-state inhalation method: In the steady-state inhalation method, 10 min PET scans are carried out after 10 min of oxygen-15-labeled $C^{15}O_2$ and $^{15}O_2$ inhalation allowing the signal to build-up to a steady-state, and a 3 min scan is done after a short period of $C^{15}O$ inhalation. Arterial blood sampling is done during each scan so that cerebral blood flow, oxygen metabolism, and cerebral blood volume can be calculated

Column: Terry Jones; a Big Man

Together with Ter-Pogossian, Terry Jones established the theory for measuring cerebral blood flow and cerebral oxygen metabolism. He also developed the steady-state inhalation method, which was an important method for early PET. He is not only a great academic but also a great human being. He developed PET measurement systems in the Cyclotron Unit of the Royal Medical Research Centre at Hammersmith Hospital, where he daily encouraged young researchers from all over the world. In the evenings, he would take everyone to the hospital bar, where they would drink lukewarm British beer, snack on potato chips, and listen to the day's research events. When I stayed at my desk because I was busy organizing my upcoming return to Japan, he angrily said, "Hey, Kanno, aren't you skipping the most important meeting here?" I was with my family, but Terry took care of us both privately and publicly. Even though Terry is in his late seventies now, he still travels around the world in high spirits and works hard to spread his dream of developing total-body PET (see Sect. 4.4.4) and establishing systemic medicine and physiology.

6.2.4 Short Inhalation Method for Oxygen-15-Labeled Gases

Under Ter-Pogossian at the University of Washington, Marcus Raichle developed a new method for measuring cerebral oxygen metabolism after intravenous injection of oxygen-15-labeled water and short-time inhalation of oxygen-15-labeled O_2 (Raichle et al. 1983). In this method, oxygen metabolism in the brain is determined by observing the kinetics of oxygen-15-labeled water metabolized in the brain and other parts of the body. This method does not require the radioactivity concentration to reach a steady-state before measurements can be made, nor is it necessary to wait for the radioactivity to return to the baseline. So, measurement time is shortened, making repeated measurements possible. The measurement time is reduced to approximately 1 h. Because of the rapidity of this method (Fig. 6.11), repeated measurements can be made under various hemodynamic perturbations such as hypercapnia and hypocapnia. It can also be used to observe neural activation in response to some external stimulus (Mintun et al. 1984). However, the measurement technique is a little more troublesome than the steady-state inhalation method because it is necessary to continuously withdraw arterial blood to sample the arterial

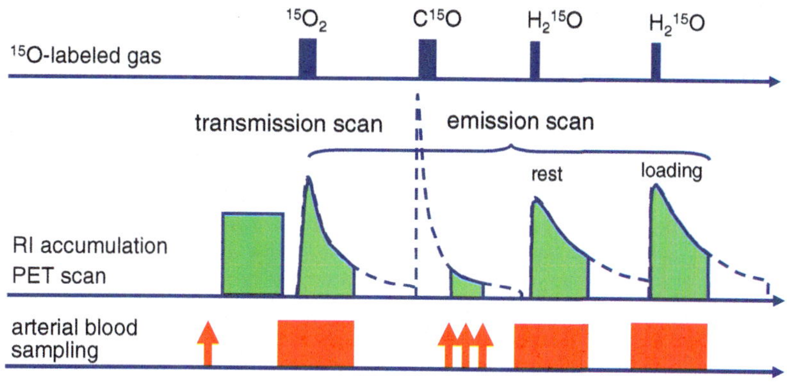

Fig. 6.11 Protocol for the oxygen-15-labeled gas short-time inhalation method: Oxygen-15-labeled O_2 and CO gases are administered sequentially at intervals of about 10 min to measure cerebral oxygen metabolism and cerebral blood volume. Multiple $H_2{}^{15}O$ injections are performed at 10 min intervals to measure cerebral blood flow during rest and necessary stimulus loading

tracer concentration. It also requires high-speed PET scanning to accurately measure the tracer concentration in time. Over the years, the method has been optimized so that one set of clinical measurements can now be performed in 15 min or less at the National Cerebral and Cardiovascular Center, Japan (Kudomi et al. 2013).

Just as Lassen's xenon-133 derivative method significantly improved Kety's laughing gas (N_2O) integral method (see Sect. 4.2.3), the short-time inhalation method was a breakthrough that dramatically expanded the possibilities of cerebral blood flow measurement with PET. The method led to important new discoveries about cerebral circulation and metabolism from the group at the University of Washington. The development of the short-time inhalation method was not a coincidence; it followed from the philosophy of Michel Ter-Pogossian and Marcus Raichle that short-time measurements are essential for assessing brain pathophysiology and neural function.

Column: Marcus Raichle's Starting Point

Marcus Raichle, who has made many contributions to brain physiology and brain science, started by developing a method to measure cerebral blood flow and metabolism using oxygen-15-labeled gas. During the time that the author was studying in London, Marcus Raichle called Terry Jones (who had already established the steady-state inhalation method with oxygen-15-labeled gas) at Hammersmith Hospital, to inquire about the normal values of cerebral blood flow and cerebral oxygen metabolism so that he could compare them with his own short-time measurement method. Later, when the author returned to the Akita Brain Institute and started measuring cerebral blood flow and cerebral oxygen metabolism using the oxygen-15-labeled gas short inhalation method, we were not sure whether our data were correct, so the author wrote a letter to Marcus Raichle. When his reply came, I was surprised to find that Marcus Raichle had sent the cerebral blood flow, cerebral oxygen metabolism, and oxygen extraction fraction data of 20–30 patients. As we were at that time a small unknown group in rural north Japan, the author was very impressed with Marcus Raichle's warm response.

6.3 Cerebral Blood Flow Measurement without RIs

This section introduces several methods for measuring cerebral blood flow that do not use RIs but are applied in clinics. In Japan, both PET and SPECT facilities must be in a strict radiation control area where RIs can be used legally. On the other hand, non-RI-based methods for measuring cerebral blood flow can be used even at medical facilities that do not have a strict radiation control area.

6.3.1 Cold Xenon Inhalation Method

In this method, non-RI (cold) xenon is inhaled instead of xenon-133 and measured with CT. Xenon, which has a high atomic number, causes a slight increase in the CT image intensity when xenon dissolves in the brain tissue. Using the same principle as the xenon-133 inhalation method, cerebral blood flow can be estimated from the brain clearance curve. However, since xenon is inhaled at a high concentration (30–35%), it is no longer a "tracer amount" (see Sect. 5.4.1) because xenon itself has an anesthetic effect like laughing gas (N_2O). Nevertheless, the cold xenon method is still used in some facilities because a radiation control area is unnecessary.

6.3.2 Arterial Spin Labeling (ASL) with MRI

A method to measure cerebral blood flow with MRI has been developed. The arterial blood is excited (spin-labeled) with an inversion pulse just before reaching the brain tissue. The excited water protons then flow within the vasculature like a diffusible tracer and eventually diffuse through the BBB into the brain tissue. MRI images of the area of interest are taken with and without spin-labeling. Since the number of spin-labeled water molecules reaching the brain tissue is in proportion to the cerebral blood flow, the cerebral blood flow can be estimated after subtracting the labeled and non-labeled images. This is called the arterial spin labeling (ASL) method (Detre et al. 1992). Unlike radioisotope (RI) methods and the methods using angiographic contrast media described in the next section, both of which rely on extrinsic tracers, the ASL method uses water in the blood itself as an intrinsic tracer. However, since the signal-to-noise ratio of the ASL method is low, an MRI system with a magnetic field strength of 3T or higher is usually needed. In addition, in the presence of cerebrovascular disease collateral circulation may invert the left-right difference on cerebral blood flow images, so care must be taken in interpreting this difference.

6.3.3 CT and MRI Contrast Agent Methods

CT and MRI are mainly used to diagnose diseases based on morphology. If an angiographic contrast agent is injected, CT and MRI may also be used to evaluate vascular deficits or leakage of the BBB in pathological conditions such as cerebral ischemia and brain tumors. At the same time, the contrast agent acts as a nondiffusible tracer, and if its behavior in the cerebral blood vessels is dynamically observed it is possible to evaluate regional cerebral blood flow from the mean transit time (MTT) (see Sect. 5.3.3). Dynamic angiographic CT or MRI measurements are widely applied in clinical practice because cerebral blood flow can be easily estimated for clinical diagnosis without the need for a radiation control area. All that is needed is an iodine contrast agent for CT or a gadolinium contrast agent for MRI. The mean transit time (MTT) of a nondiffusible tracer in normal brain is only a few seconds, however, in cases of cerebrovascular disease there may be a delay of several seconds in either or both the left and right hemispheres due to circulatory delay via collateral circulation. This can easily lead to erroneous evaluation of the left and right hemispheres, which requires great care and hence the method is less reliable than SPECT.

6.3.4 Evaluating Cerebral Blood Volume with Near Infrared Spectroscopy (NIRS)

As the cost of lasers decreased, the signal-to-noise ratio of detectors and other technology necessary for NIRS measurements improved dramatically. Consequently, NIRS has developed rapidly as a technology to create images of oxygenated and deoxygenated hemoglobin from outside the cranium (Fig. 6.12). However, NIRS does not measure cerebral blood flow so it cannot be used for disease diagnosis. It is instead used as a method to detect changes in the ratio of oxygenated hemoglobin concentration to deoxygenated hemoglobin concentration during the excess cerebral blood flow stimulated by brain activation.

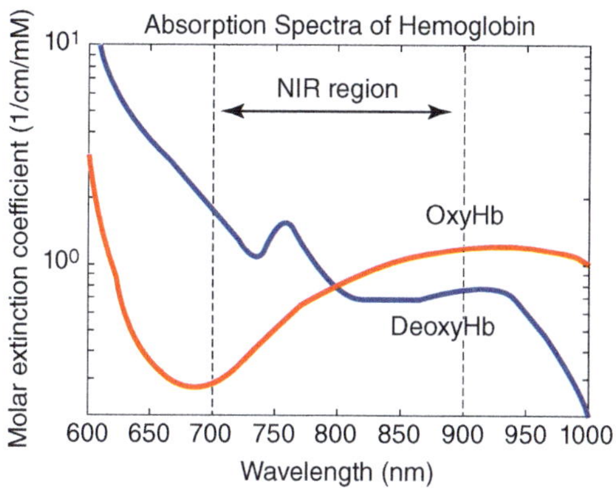

Fig. 6.12 Absorption coefficients of oxy- and deoxy-hemoglobin: Since the absorption coefficients of oxyhemoglobin (HbO: red) and deoxyhemoglobin (Hb: blue) differ along to the wavelength, the oxygenation rate of blood can be determined from the ratio of HbO to Hb at each wavelength. (Courtesy of Yamada, UEC, Tokyo)

The NIRS method has advantages and limitations. There are scalp and skull obstacles between the probe and the cortical surface that scatter the incident and reflected near-infrared waves, resulting in very low spatial resolution. The spatial resolution is mostly limited by the fact that the signal is affected by the scalp and the hair on the subject's head. On the other hand, the advantage of NIRS is that it only requires a near-infrared laser source, a pair of optical fibers to transmit and receive the light, and a near-infrared light detector. A typical NIRS device is portable and is easily placed on the subject's head, and measurements can be made in real time. This makes it possible to design brain activation tasks that are not restricted to the bedside and thus provides a highly flexible method for measuring brain function (Fig. 6.13).

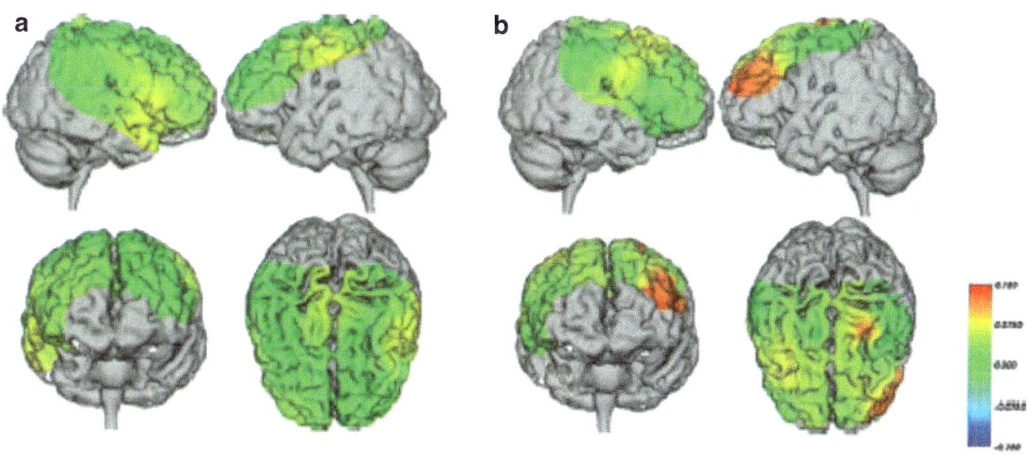

Fig. 6.13 Near infrared spectroscopy (NIRS) images: A. Control image of HbO concentration (walking speed 4 km/h). B: Image of HbO concentration during obstacle avoidance gait (walking speed 4 km/h). After subtracting image A from image B, there was an increase in HbO concentration in the right premotor cortex and the left prefrontal cortex (near dorsolateral cortex). (Courtesy of Miyai I, Morinomiya Hospital, Osaka)

6.3.5 Measuring Bulk Flow with Ultrasound

Ultrasound (US) is a method that can measure bulk flow but not perfusion flow (see Sect. 5.1.2). Due to the Doppler shift, US can be used to estimate the bulk flow velocity by placing the probe parallel to the direction of the vascular path. At the bedside, the flow velocity of the middle cerebral artery can be estimated from the Doppler shift by placing the probe over the temple area, but care must be taken to make sure the blood vessel is in the same orientation as the ultrasound probe. Even under various conditions, such as a left-right vascular deficit and CO_2 inhalation, that may limit the effectiveness of this method, it is useful because it can be easily used to evaluate relative changes in blood flow velocity at the bedside.

6.4 Cerebral Blood Flow Measurement in Animal Experiments

Accurate measurement of cerebral blood flow is necessary not only for clinical medicine but also for basic research using animals. Such measurements have been performed even before Kety's time. New technology has enabled the development of various practical methods.

6.4.1 Hydrogen Clearance Method

Since hydrogen gas is a diffusible inert gas, it can be used as a diffusible tracer in the same way as xenon-133. When hydrogen is inhaled for a short period of time, several tens of seconds, it immediately diffuses and saturates the brain tissue. When hydrogen inhalation is stopped, the hydrogen is washed out by fresh blood. Measurement of the hydrogen brain clearance curve can be used to determine cerebral blood flow (Aukland et al. 1964). The hydrogen concentration in the brain is measured with a platinum electrode surgically inserted into the area of interest within the brain prior to the experiment. If multiple platinum electrodes are used, hydrogen cerebral clearance curves can be measured simultaneously for multiple regions, and the cerebral blood flow at each electrode position can be measured. If the electrode insertion position is in a region where there is a mix of gray matter and white matter, the brain clearance curve will be the sum of two exponential decays. If biexponential fitting of the data is performed, the cerebral blood flow of each component can be estimated.

6.4.2 Autoradiography (ARG) Method

In the autoradiography (ARG) method, tracers labeled with RIs that emit beta rays (e.g. carbon-14, iodine-125, and tritium) are administered to experimental animals, and the brains are removed at specific times according to the tracer analysis model. Since the ARG method requires multiple brain samples at each measurement time point, the number of animals used in the experiment is usually large. Sakurada developed a method to quantify cerebral blood flow using carbon-14-labeled iodide antipyrine (IAP) (Sakurada et al. 1978). The ARG method with carbon-14-labeled IAP was the only method that could measure quantitative cerebral blood flow at that time, and its use in animal experiments spread explosively. Sokoloff's carbon-14-labeled DG method for quantitative measurement of glucose metabolism was also performed using the ARG method. Thus, cerebral glucose metabolism and cerebral blood flow can now be measured with the same ARG technique.

6.4.3 Cerebral Blood Flow Measurement Using Light Transmission

This is a method for quantifying cerebral blood flow developed by Minoru Tomita (Fig. 6.14) and his colleagues at Keio University, Japan. The experimental preparation is complicated. A small light bulb is implanted under the cerebral cortex of the animal in the region to be measured, and a

Fig. 6.14 Minoru Tomita (1934–2010)

photodiode fixed on the cortical surface receives light transmission from the bulb. The experiment is performed after allowing the animal sufficient time to recover from preparation surgery. When saline solution is injected in a bolus through the internal carotid artery, concentration changes in the blood occur after a few seconds and perturb the light transmitted to the photodiode. The mean transit time MTT of the saline solution can be calculated from the transit of the transparent saline solution. This method is excellent for quantitative measurements of cerebral blood flow in animal experiments. Moreover, the cerebral blood flow is localized, and repeated measurements under various physiological conditions can be performed by injecting fresh saline (Tomita et al. 1978).

6.4.4 Laser-Doppler Flowmetry (LDF)

The decrease in cost of lasers enabled the development of a method to evaluate cerebral blood flow using a near-infrared laser. Laser Doppler flowmetry (LDF) was developed in the 1980s as a method to evaluate cerebral blood flow at a microscopic level. It has become an essential method for monitoring blood flow during animal experiments. Microvascular cerebral blood flow is measured by irradiating the red blood cells moving in capillaries with a near-infrared laser light, and then recording the reflected light and Doppler shift to estimate red blood cell density and velocity. Normally, the irradiating fiber and the receiving fiber are sealed together in a thin stainless-steel tube. The result of the measurement is equivalent to the average cerebral blood flow in an area of approximately 1 mm^3 just below the probe (Fig. 6.15). The result is a relative cerebral blood flow value, but if the probe is fixed, changes in cerebral blood flow can be accurately tracked. As well as animal experiments, LDF is used on humans during neurosurgery to monitor changes in cerebral blood flow in the brain tissue, although it only measures a single localized point directly under the fiber probe.

6.4.5 Laser Speckle Flow Imaging (LSFI)

Laser speckle flow imaging (LSFI) measures cerebral blood flow in a two-dimensional plane using a CCD camera that records the near-infrared laser light scattered off red blood cells. As with LDF, it cannot measure absolute blood flow, but it does accurately monitor temporal and relative changes to the cerebral blood flow across a two-dimensional region (Fig. 6.16). LSFI has become a useful method for observing the two-dimensional distribution of cerebral blood flow over a wide area of the brain surface in various animal experiments. It is a useful method for monitoring cerebral blood flow, especially in rodent experiments, where a relatively wide distribution of cerebral blood flow over a few millimeters squared in the brain is often needed experimentally.

6.4.6 Confocal Microscopy and Two-Photon Microscopy

In the 2000s, *in vivo* fluorescence microscopy, which scans a relatively high-power laser beam at high speed across the item under observation, became available for practical use. Confocal microscopy can now be used on animals that have been genetically engineered to express fluo-

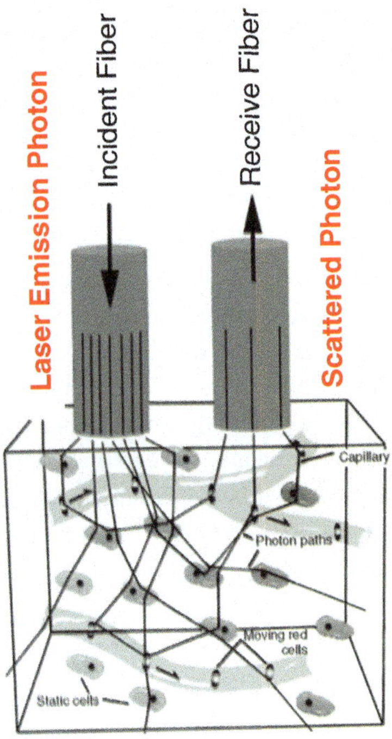

Fig. 6.15 Principle of laser-Doppler flowmetry (LDF): When infrared laser light irradiates capillaries through the incident fiber, the Doppler shift of the scattered light is detected via the receive fiber. By analyzing the spectra and intensity of the received light, the velocity and volume of red blood cells are calculated

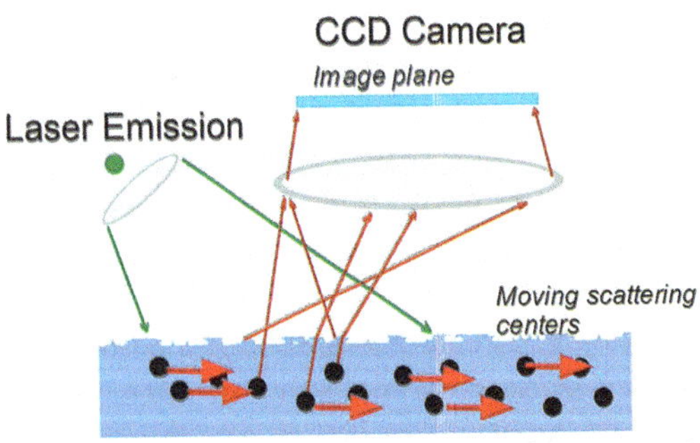

Fig. 6.16 Principle of laser speckle flow imaging (LSFI): A laser beam is scanned over the brain surface and reflected light is measured by a CCD camera. After image processing, the speed of the red blood cell movement can be obtained. The signal also reflects cerebral blood flow

rescent molecules. The brain surface can be observed down to a depth of approximately 100 μm by scanning brain surface tissue with a laser of wavelength that excites the fluorescent molecules. The fluorescent material transitions to an electronically excited state and fluorescence light is emitted when the electrons return to the ground state (Fig. 6.17).

Two-photon excitation is the simultaneous absorption of two photons of identical frequencies to excite a molecule from the ground state to a higher energy, most commonly an excited electronic state. In contrast to the one-photon excitation of confocal microscopy, there are two-photon excitation processes where two photons are simultaneously absorbed and increase their energy (and shorten the wavelength) by a factor of two, and three-photon excitation processes where three photons are simultaneously absorbed and increase their energy (and shorten the wavelength) by a factor of three. In two-photon excitation, the wavelength of the incident laser light is in the near-infrared region (which is double the wavelength of the light required to excite fluorescence), making it possible to measure depths in the brain down to approximately 1000 μm. In vivo measurement of cerebral vasculature (surface and penetrating vessels) and capillaries during neural activation is now possible with a resolution of less than 1 μm (Fig. 6.18). By observing the vascular

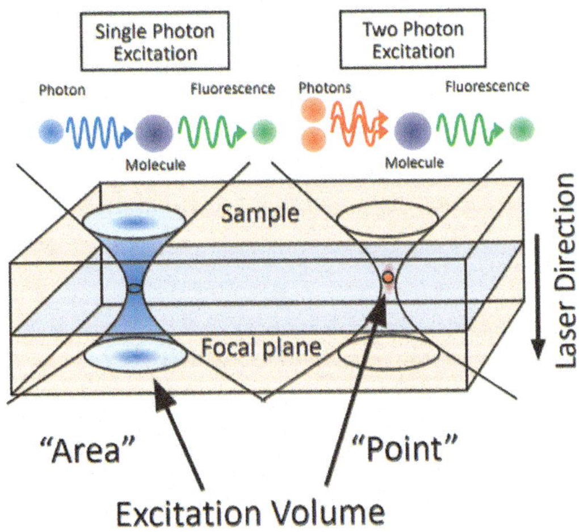

Fig. 6.17 Principle of two-photon laser microscopy: Two-photon excitation is the simultaneous absorption of two photons of identical frequencies to excite a molecule from the ground state to a higher energy level. This microscope can excite fluorescence at twice the wavelength of the irradiated light through a two-photon absorption process. If infrared light is used as the excitation, observation of the brain surface to a depth of about 1 mm is possible. (Courtesy of Kawakami R and Nemoto T, 2019)

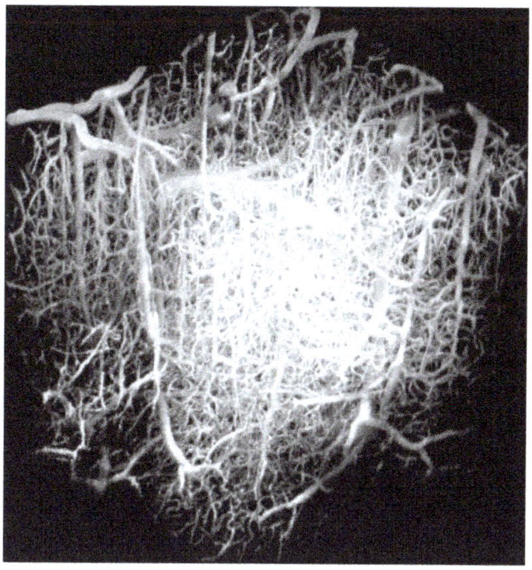

Fig. 6.18 Capillary network of the rat cerebral cortex observed with two-photon microscopy: The XYZ scales are 900 × 900 × 800 μm

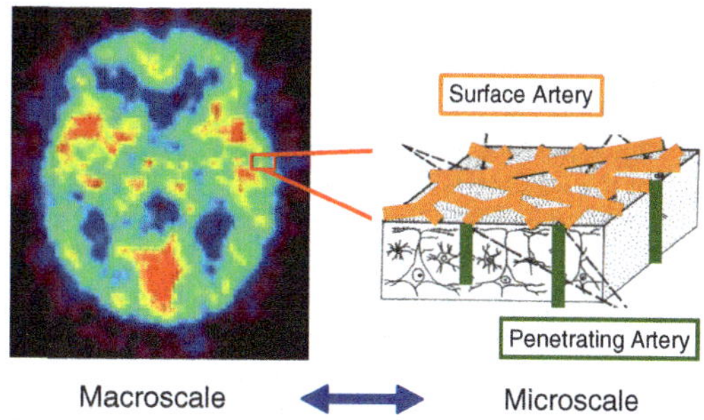

Fig. 6.19 Demonstration of the relationship between macroscopy and microscopy: Macroscopic measurements with PET and microscopic measurements with two-photon microscopy images are linked. Regional cerebral blood flow changes in PET can be confirmed by observing hemodynamic changes in microvessels with two-photon microscopy

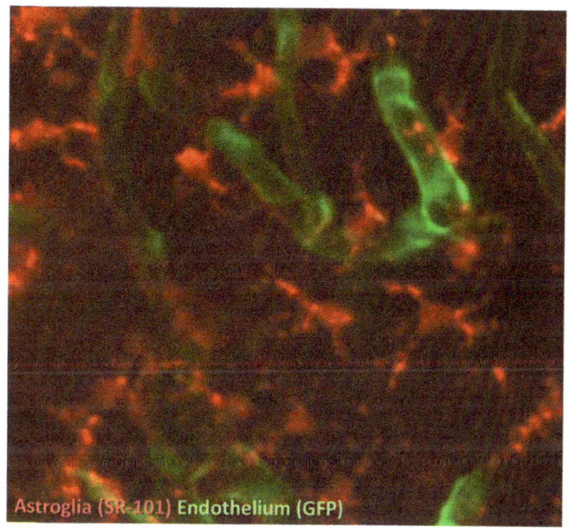

Fig. 6.20 Simultaneous imaging of multiple cells with two-photon microscopy: Simultaneous measurement of astrocytes (red), stained by peritoneal administration of SR101, and vascular endothelial cells (green), which express GFP, in Tie2 mice

network in three dimensions, real-time changes in vessel diameter can be made for vessels both on the cortical surface and penetrating the brain parenchyma. Importantly, during brain activation it is possible to measure the behavior of multiple vessels independently and simultaneously in the activated area (Fig. 6.19).

The key technologies for these microscopy methods are genetic engineering and the development of chemical substances that emit fluorescence when excited by laser light. Recent genetic engineering techniques have made it easy to express fluorescent proteins such as GFP (green fluorescence protein) and RFP (red fluorescence protein) in specific cells of mice. These proteins, along with calcium, are associated with neuron firing. Simultaneous measurement of these proteins and calcium-ion fluorescence associated with neuron firing is now possible. By combining these genetic modification technologies with the technology to express fluorescent proteins independently in the component cells of the NVU, it is possible to observe the independent behavior of each cell. For example, using Tie2 mice expressing GFP (green) in endothelial cells, changes in the diameter of capillary vessels and the location of astrocytes stained with peritoneal SR101 (red) have been imaged (Fig. 6.20). Changes in the diameter of arterioles, capillaries and venules, and even pericytes in combination with physiological stimuli and various drug loads, allow changes in cerebral blood flow to be observed. The three-dimensional structure of the capillaries and the movement of microglia, pericytes, astrocytes, and endothelial cells can be continuously observed in real time. These techniques are bringing us closer to understanding the mechanism behind neurovascular coupling (NVC).

References

Aukland K, Bower BF, Berliner RW. Measurement of local blood flow with hydrogen gas. Circ Res. 1964;14(2):164–87.

Detre JA, Leigh JS, Williams DS, Koretsky AP. Perfusion imaging. Magn Reson Med. 1992;23(1):37–45.

Frackowiak RSJ, Lenzi GL, et al. Quantitative measurement of regional cerebral blood flow and oxygen metabolism in man using ^{15}O and positron emission tomography: theory, procedure, and normal values. J Comput Assist Tomogr. 1980;4(6):727–36.

Iida H, Ito H, et al. A method to quantitate cerebral blood flow using a rotating gamma camera and iodine-123 iodoamphetamine with one blood sampling. Eur J Nucl Med. 1994;21(10):1072–84.

Ito H, Inoue K, et al. Database of normal human cerebral blood flow measured by SPECT: I. Comparison between I-123-IMP, Tc-99m-HMPAO, and Tc-99m-ECD as referred with O-15 labeled water PET and voxel-based morphometry. Ann Nucl Med. 2006;20(2):131–8.

Jones T, Chesler DA, Ter-Pogossian MM. The continuous inhalation of oxygen-15 for assessing regional oxygen extraction in the brain of man. Br J Radiol. 1976;49(580):339–43.

Kanno I, Lassen NA. Two methods for calculating regional cerebral blood flow from emission computed tomography of inert gas concentrations. J Comput Assist Tomogr. 1979;3(1):71–6.

Kawakami R, Nemoto T. In vivo imaging of all cortical layers and hippocampal, CA1 pyramidal cells by two-photon excitation microscopy. In: Kao F-J, Keiser G, Gogoi A, editors. Advanced optical methods for brain images. Singapore: Springer Nature; 2019.

Kim KM, Watabe H, et al. Quantitative mapping of basal and vasoreactive cerebral blood flow using split-dose 123I-iodoamphetamine and single photon emission computed tomography. NeuroImage. 2006;33(4):1126–35.

Kudomi N, Hirano Y, et al. Rapid quantitative CBF and CMRO(2) measurements from a single PET scan with sequential administration of dual (15)O-labeled tracers. J Cereb Blood Flow Metab. 2013;33(3):440–8.

Lammertsma AA, Jones T. Correction for the presence of intravascular oxygen-15 in the steady-state technique for measuring regional oxygen extraction ratio in the brain: 1. Description of the method. J Cereb Blood Flow Metab. 1983;3(4):416–24.

Lassen NA, Ingvar DH. The blood flow of the cerebral cortex determined by radioactive krypton. Experientia. 1961;17:42–3.

Mintun MA, Raichle ME, et al. Brain oxygen utilization measured with O-15 radiotracers and positron emission tomography. J Nucl Med. 1984;25(2):177–87.

Obrist WD, Thompson HK Jr, et al. Determination of regional cerebral blood flow by inhalation of 133-xenon. Circ Res. 1967;20:124–35.

Raichle ME, et al. Brain blood flow measured with intravenous $H_2(15)O$. II. Implementation and validation. J Nucl Med. 1983;24(9):790–8.

Sakurada O, Kennedy C, et al. Measurement of local cerebral blood flow with iodo [14C] antipyrine. Am J Phys. 1978;234(1):H59–66.

Ter-Pogossian MM, Eichling JO, et al. The measure in vivo of regional cerebral oxygen utilization by means of oxyhemoglobin labeled with radioactive oxygen-15. J Clin Invest. 1970;49(2):381–91.

Tomita M, Gotoh F, et al. Photoelectric method for estimating hemodynamic changes in regional cerebral tissue. Am J Phys. 1978;235(1):H56–63.

West JB, Dollery CT, et al. The use of radioactive carbon dioxide to measure regional blood flow in the lungs of patients with pulmonary disease. J Clin Invest. 1961;40:1–12.

Yamamoto YL, Thompson CJ, et al. Dynamic positron emission tomography for study of cerebral hemodynamics in a cross section of the head using positron-emitting 68Ga-EDTA and 77Kr. J Comput Assist Tomogr. 1977;1(1):43–56.

7 Cerebral Blood Flow Supporting the Brain

Abstract

This chapter describes how cerebral blood flow supports the brain in health and disease. Since the brain is so important it is protected by a system that maintains a stable cerebral blood flow supply. In the healthy brain, cerebral blood flow is regulated by four basic regulating mechanisms: blood pressure regulation (also known as autoregulation), chemical regulation, neurogenic regulation, and metabolic regulation. However, in diseased brain, vascular deficit of a proximal area causes a reduction in cerebral perfusion pressure (CPP) which reduces cerebral blood flow and results in misery perfusion and later luxury perfusion. In the chronic phase, circulation reserve and the metabolic reserve are decreased. With changes in pathology, such as neurodegeneration or accumulation of abnormal protein (e.g. beta-amyloid or tau protein) in distal areas, cerebral blood flow in those areas is naturally reduced.

Keywords

Autoregulation · Chemical regulation · Neurogenic regulation · Metabolic regulation · Misery perfusion · Luxury perfusion · Circulation reserve · Metabolic reserve

7.1 Regulation of Cerebral Blood Flow

This section summarizes the basic regulatory mechanisms for cerebral blood flow. The macroscopic environment for cerebral blood flow in the cranium will be described first, and then the microscopic environment will follow.

7.1.1 Intracranial Regulation of Cerebral Blood Flow

The arteries are always under internal pressure. This pressure pulsates. The average of the highest and lowest blood pressures of one beat of the heart is called the mean arterial blood pressure (MABP). MABP is the driving force that pushes blood from upstream to downstream, and cerebral blood flow is supplied intracranially via the internal carotid artery and the vertebral artery. It is believed that 60% of the vascular resistance is in the extracranial vessels, and the remaining 40% is in the intracranial vessels (Stromberg and Fox 1972). The intracranium is a space enclosed by the skull and filled with brain tissue, blood, and spinal fluid in approximate proportions of 85%, 5%, and 10% by volume, respectively. The contents of the intracranium compress cere-

I. Kanno, *What Cerebral Blood Flow Tells Us*, https://doi.org/10.1007/978-981-97-6382-5_7

bral blood vessels and this is called intracranial pressure (ICP: approximately 6–12 mmHg), which acts as a factor to reduce the cerebral perfusion pressure (CPP) that drives cerebral blood flow within the skull. In other words, CPP = MABP − ICP. Similar to Ohm's law in electrical circuits, the actual cerebral blood flow (f) is proportional to cerebral perfusion pressure CPP (p) and inversely proportional to vascular resistance (r), $f = p/r$.

In addition to blood vessel diameter, the viscosity of blood also has a significant effect on vascular resistance. Blood viscosity is essentially a systemic factor that contributes to vascular resistance. On the other hand, if vascular resistance is constant, cerebral blood flow will change in response to changes in MABP. Since MABP changes with the posture of the trunk and various other factors, the cerebrovascular autoregulation mechanism described in the next section will function intracranially to maintain constant cerebral blood flow by changing the arteriolar diameter in opposition to the local cerebral perfusion pressure. In this way, cerebral blood flow is maintained at a constant level.

7.1.2 Regulation in the Microvasculature

In blood vessels smaller than 100 μm in diameter, the flow of blood, consisting of plasma and blood cell components, is not Newtonian. Blood is composed of solid components (erythrocytes, leukocytes, platelets) and liquid components (plasma: an aqueous solution of electrolytes and organic matter) and has non-Newtonian viscosity. In vessels below 100 μm in diameter, the viscosity is reduced due to the presence of red blood cells, which account for most of the solid components. This is called the Fåhleus-Lindqvist effect. Simulation results show that if the red blood cells were absent, but hemoglobin pigment was dissolved in the blood in the same concentration, then the burden on the heart would increase by 30–40% (Snyder 1973).

Red blood cells flow through blood vessels in a fluid called plasma, and their shape and flexibility allow them to circulate through capillaries in the lungs and tissues. However, if the normally disk-shaped red blood cells are deformed into a spherical or other irregular shape due to factors such as a decrease in osmotic plasma pressure, a change in the cell membrane, the effect of drugs, or various diseases, then the ability of red blood cells to alter their shape is reduced, making it difficult for them to pass through capillaries. In addition, there is a genetic disorder that causes red blood cells to have an abnormal shape. This is a disorder in which some red blood cells are deformed into a sickle shape. Since sickle cells are harder than normal red blood cells, they cannot deform smoothly to pass through small capillaries. They also break easily, resulting in impaired blood flow and O_2 transport.

7.2 Regulation of Cerebral Blood Flow in Healthy Brains

This section introduces the regulatory mechanisms for cerebral blood flow in a healthy brain. Lassen identified four basic factors in the regulation of cerebral blood flow: perfusion pressure regulation, chemical regulation, metabolic regulation, and neurogenic regulation (Fig. 7.1).

7.2.1 Perfusion Pressure Regulation (Autoregulation)

The perfusion pressure of the brain changes depending on blood pressure. Blood pressure fluctuates with changes in body posture and other factors. In the head and neck region, which is located on the top of the trunk, blood pressure to the head fluctuates when the body bends over or stands up. The cerebral perfusion pressure (CPP) in cerebral blood vessels also fluctuates accordingly. If the cerebral blood flow changes from moment to moment depending on the perfusion pressure, the supply of nutrient substances to the

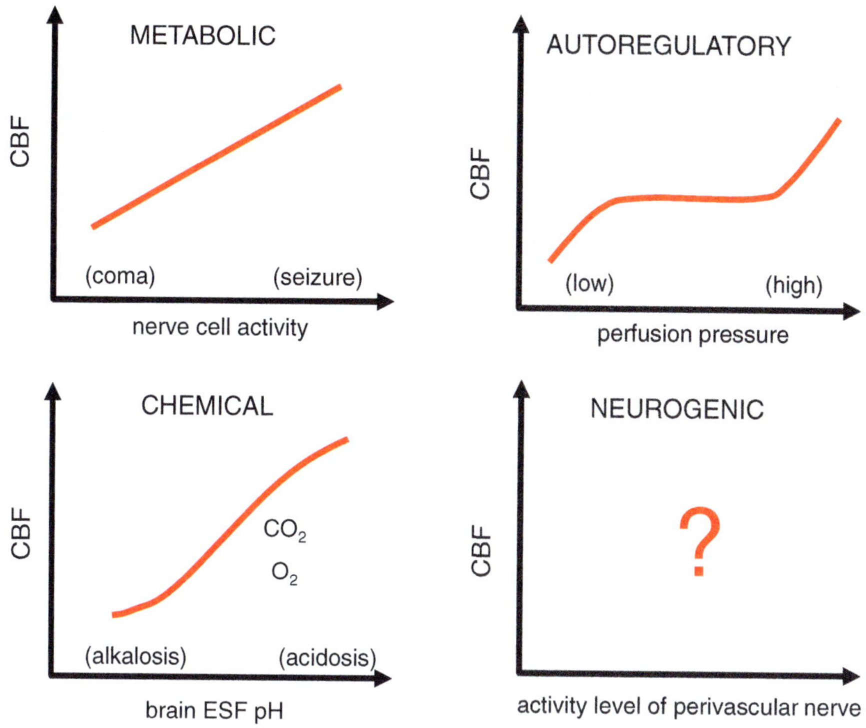

Fig. 7.1 Classical theory of cerebral blood flow (CBF) regulation: Lassen proposed that CBF is regulated based on four physiological mechanisms, i.e., metabolic regulation, autoregulation, chemical regulation, and neurogenic regulation. (Lassen NA. Circ Res. 1974)

brain will become unstable, so a safety mechanism called the autoregulation mechanism is provided to maintain a constant cerebral blood flow in response to blood pressure fluctuations. It is the most basic regulation mechanism to maintain a steady cerebral blood flow. When CPP rises, cerebral blood vessels contract to increase vascular resistance, and when CPP falls, cerebral blood vessels dilate to lower vascular resistance to maintain a constant cerebral blood flow. If the CPP is within a certain range, the cerebral blood flow will be maintained at a constant level (Fig. 7.2). The function of vascular smooth muscle to relax or contract vessel diameter in response to changes in luminal pressure is called myogenic regulation. The autoregulation mechanism of myocardial blood vessels has been studied in detail and is thought to work in the following series of actions: increase in perfusion pressure → increased load on arterial wall → contraction of small arterioles → increase in blood flow resistance → decrease in cerebral blood flow. A mechanism like that for myogenic regulation has been postulated for cerebral vessels. Gotoh et al. proposed the dual regulation theory where the autoregulation mechanism is played by blood vessels larger than 50 μm (Gotoh et al. 1975).

The autoregulation range in the brain is confined between the lower and upper limits of the CPP (approximately 50–150 mmHg). When CPP falls below the lower limit of the autoregulation range, the diameter of the cerebral blood vessel cannot relax (dilate) any further, and cerebral blood flow decreases. On the other hand, if systemic blood pressure rises abnormally, such as when straining in the toilet or having sexual intercourse, it exceeds the upper limit of the autoregulation range, and cerebral blood flow increases because cerebral blood vessel diameter cannot contract sufficiently to compensate. If the

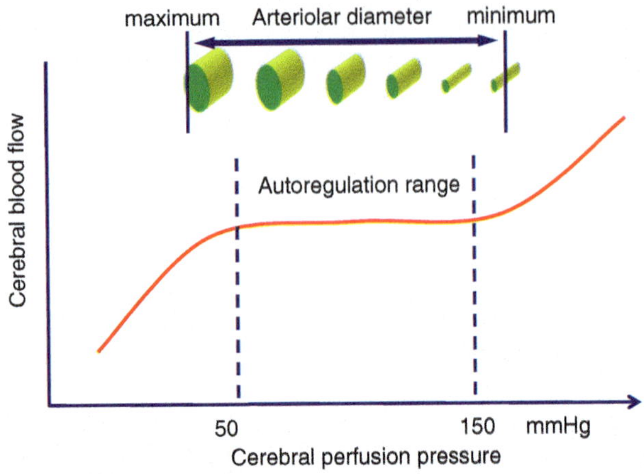

Fig. 7.2 Relationship between cerebral perfusion pressure (CPP) and arteriolar diameter: While CPP is within the autoregulation range, the arteriolar diameter increases (or decreases) in response to decreased (or increased) perfusion pressure. That is, the arteriolar diameter contracts or dilates as the CPP increases or decreases, respectively, so that the cerebral blood flow can be kept constant

blood pressure remains out of the range of the autoregulation range, the brain tissue will be in a very dangerous state.

7.2.2 Chemical Regulation

Chemical regulation may contribute to regulation locally or across the whole brain. If the partial pressure of CO_2 in arterial blood ($PaCO_2$) increases due to CO_2 inhalation, CO_2 diffuses from the blood into the tissue surrounding the blood vessel and reacts to become bicarbonate ions (HCO_3^-), and the protons (hydrogen ions: H^+) produced in this reaction act directly on the smooth muscle to cause vascular dilation (Fig. 7.3) (Lassen 1968). Conversely, hyperventilation lowers $PaCO_2$ and decreases hydrogen ions, causing smooth muscle to contract and blood vessel diameter to reduce. During local cerebral ischemia the brain tissue becomes acidotic and blood vessels relax so that cerebral blood flow increases due to hydrogen ions in the perivascular area. The equivalent of elevated $PaCO_2$ can also be caused by the administration of acetazolamide (ACZ). ACZ is a carbonic acid dehydratase inhibitor, but the CO_3^- ion produced by aerobic metabolism of glucose in the brain is not discharged as CO_2 into blood

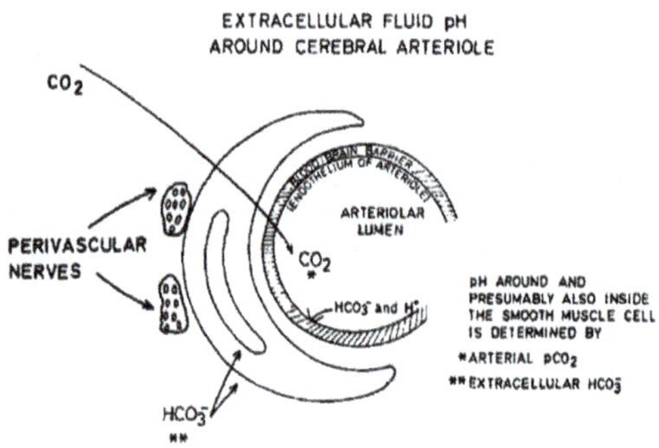

Fig. 7.3 Chemical regulation of cerebral arteries: When the partial pressure of CO_2 in arterial blood is increased, the pH of peri-arterial smooth muscle is lowered, causing the muscle to relax. (Lassen NA Scand J Clin Lab Invest 1968)

vessels due to inhibition of dehydratase action by ACZ, and the concentrations of CO_3^- and H^+ around cerebral blood vessels increase, resulting in dilatation of cerebral blood vessels as in the mechanism described above.

There are also chemical vasoactive factors that act locally. One of these, nitric oxide (NO) was discovered in the 1980s to be a powerful vasodilator and was initially called endothelium-derived relaxing factor (EDRF). Later, in 1990, three types of NO synthase (NOS) were purified from nerves, endothelium, and macrophages (called nNOS, eNOS, and iNOS, respectively), and inhibitors for these enzymes have also been discovered. However, studies have shown that inhibition of NO synthesis by NO synthase inhibitors alone does not completely inhibit vasodilation. Even in the most basic CO_2-responsive state, only 50% of the increase in cerebral blood flow can be suppressed. NO is a type of free radical and, as well as vasodilation, is known to cause various physiological effects.

Carbon monoxide (CO) is another chemical factor that has attracted attention as a vasoconstrictor. The mechanism by which CO is produced in vivo is still not fully understood. In addition to the gaseous substances CO and NO, several chemical mediators, such as angiotensin II (AT2), prostaglandin I2 (PGI2), and cyclooxygenase 2 (COX–2), have been implicated in the dilation and constriction of local cerebral blood vessels.

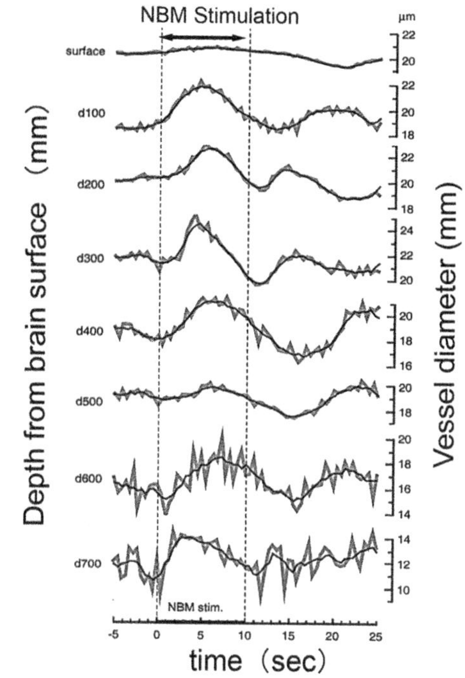

Fig. 7.4 Neurogenic regulation: Dynamics of the cortical vessel diameter measured using two-photon microscopy during stimulation of the Nucleus basalis Meynert (NBM). Only the diameter of the penetrating vessels responds to NBM stimulation. The diameter of surface vessels is unaffected. (Modified from Hotta H, et al., J Cereb Blood Flow Metab 2013)

7.2.3 Neurogenic Regulation

The cerebral arteries that supply blood flow to the brain are connected to various sympathetic and parasympathetic nerve endings, and cerebrovascular regulation via these nerves is called neurogenic regulation. It is known that sympathetic nerve stimulation causes local changes in the smooth muscle tone of cerebral blood vessels. The presence of sympathetic nerve terminals in peripheral arterioles has been known for quite some time, but their role is not fully understood. More recently, it has been found that parasympathetic nerve terminals are also attached to the arterioles. Lassen marked the relationship between cerebral blood flow and sympathetic nerves with a question mark in his 1974 review (Fig. 7.1) (Lassen 1974), and in subsequent reviews he drew a single horizontal line to indicate that there was little involvement and concluded that "neural regulation is hardly involved in the regulation of cerebral blood flow".

Recently, as a variety of stimulation methods and detailed measurements have become available, examples have been presented that require us to rethink Lassen's view of neural regulation. For example, it has been observed with two-photon excitation microscopy that stimulation of the nucleus basalis of Meynert (NBM) causes the penetrating arteries to relax (dilate) within seconds, but the surface artery does not because the superficial and penetrating vessels of the cortex are innervated differently (Fig. 7.4) (Hotta et al. 2013). The validity of conventional neural regulation is being re-examined using two-photon excitation microscopy, and conventional textbooks are being rewritten.

7.2.4 Metabolic Regulation

Since the era of Mosso and Roy & Sherrington, it has been widely recognized that cerebral blood flow changes with brain activity. This is called metabolic regulation. Metabolic regulation is the action of increasing cerebral blood flow in proportion to metabolic activity. Lassen's hypothesis for metabolic regulation was that cerebral metabolic activity produces CO_2, which dilates cerebral blood vessels in the active area. Under the hypothesis it would be expected that in the extreme case of epilepsy, where all neurons in a certain area become excited, the cerebral blood flow in that area also increases to the maximum. Similarly, in the opposite extreme during coma, where neurons are dormant, it would be expected that cerebral blood flow is substantially reduced. These expectations were reasonable in the days when only the average cerebral blood flow could be measured macroscopically. However, when microscopic measurement, or even direct measurement of individual blood vessels, with a laser Doppler flowmeter became practical, the expectations were found to be unreasonable. The time from peripheral nerve stimulation to cerebral blood flow response in the cerebral cortex is less than 0.5 s, which is too fast to be explained as metabolic CO_2-induced dilatation. In addition, cerebral blood flow increases in response to stimulation even when it has already been elevated by hypercapnia. This suggests the involvement of mediators or direct neuronal dilators independent of hypercapnia. Elucidating the mechanism regulating cerebral blood flow response to peripheral stimulation, or neurovascular coupling (NVC), is a major focus of current ongoing research (see Chap. 8).

7.2.5 Why Is the Oxygen Extraction Fraction 0.4?

The oxygen extraction fraction (OEF) is one of the most important parameters in PET measurements of cerebral metabolism. OEF indicates the balance between oxygen supply and consumption and is a very useful index to evaluate the pathophysiology during cerebral ischemia. Measurement of cerebral circulation and metabolism in healthy adults at rest shows that cerebral blood flow (CBF) and cerebral metabolic rate of oxygen (CMRO2) are balanced in equilibrium everywhere in the brain, and OEF (= 0.4) is uniform across the brain (Fig. 7.5). Since OEF is the ratio of cerebral oxygen metabolism to cerebral oxygen supply (= CBF × oxygen content), the inverse of the OEF implies that cerebral oxygen supply is 2.5 times greater than cerebral oxygen metabolism in the resting state. In other words,

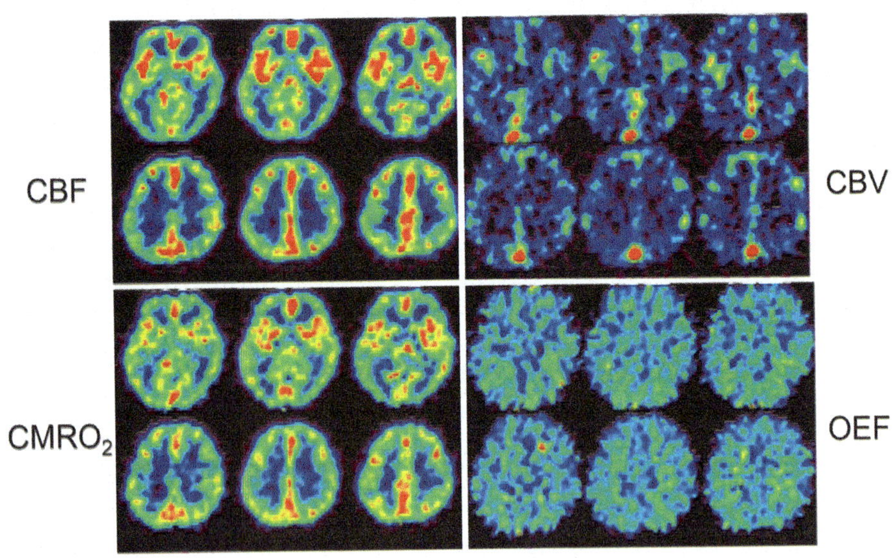

Fig. 7.5 PET measurements of the cerebral circulation and oxygen metabolism of a healthy human at rest: Cerebral blood flow (CBF), cerebral metabolic rate of oxygen (CMRO2), cerebral blood volume (CBV), and oxygen extraction fraction (OEF) are shown

cerebral blood flow supplies 2.5 times as much oxygen as is used by the brain for oxygen metabolism. The supply of oxygen by the CBF is extremely luxurious in comparison to oxygen demand. It is thought that excessive cerebral oxygen supply at the macroscopic level is necessary to maintain the partial pressure of O_2 in brain tissue near the capillaries at the microscopic level, but why the ratio of cerebral oxygen supply to cerebral oxygen consumption is balanced at 2.5 remains a mystery.

7.3 Cerebral Blood Flow During Problems on the Proximal Side

Until the 1970s, stroke was the leading cause of death among Japanese people. However, powerful campaigns to reduce salt intake and improve nutrition have reduced the occurrence of arteriosclerosis and cerebral hemorrhage caused by high blood pressure. In the early 2000s, the mortality from severe cerebrovascular disorders and strokes continued to decrease, falling below that of cancer, heart disease, and, more recently, pneumonia, to the fourth leading cause of death. The incidence of cerebrovascular disease or stroke, however, has not decreased much. The number of patients who experience the after-effects of brain function damage from cerebrovascular disease and whose quality of life (QOL) has reduced is still increasing. There are many cases of motor and sensory paralysis, language impairment, loss of autonomy, mental depression, and even dementia, and when considering the after-effects, cerebrovascular disease is the most important disease to overcome in an aging society. The loss to society caused by this disease has become one of Japan's biggest social problems. Recently, due to adverse changes in dietary habits, there has been a gradual increase in the number of cases where cerebrovascular disorders progress due to the narrowing of cerebral blood vessels caused by atherosclerotic degeneration and insufficient circulation in peripheral blood vessels. It is therefore important for clinical diagnosis to properly evaluate chronic abnormalities in cerebral blood flow and metabolism.

Column: Stroke Incidence in Akita

In Akita, having a stroke is called "ataru", meaning "hit the head". Strokes were common in Akita Prefecture in the 1960s. Cerebral hemorrhage was a major social problem that often killed the family mainstay in their 30s or 40s. The local governor of Akita Prefecture established the Akita Cerebrovascular Research Center (Akita Brain Institute) in 1968 with the cooperation of Tohoku University School of Medicine. They quickly established a new stroke treatment method called the "Akita method," in which patients with strokes are transported by ambulance and treated surgically during the acute stage. At the same time, they launched a campaign to change the high salt and low protein diet that was the main cause of strokes. As a result, the mortality rate for cerebral hemorrhage has been reduced. However, the mortality rate of cerebral infarction did not decrease, so the mortality rate of stroke continued to compete for the top cause of death in Akita until approximately 1980. Gradually, however, the results of the dietary campaign became pervasive, the stroke mortality rate decreased, and the disgrace of being number one in Japan for strokes was disposed of. In 1983, the Akita Brain Institute invested in a wealth of research equipment to eradicate strokes; it established a PET center with a compact cyclotron, and it was the leading Japanese center for stroke diagnosis and treatment research until the 2000s.

7.3.1 Misery Perfusion and Luxury Perfusion

We need to reconsider the pathophysiology of cerebrovascular disease from the perspective of the balance between the supply of and demand for oxygen. The oxygen extraction fraction (OEF) is an important parameter that reflects the balance between cerebral oxygen supply and cerebral oxygen metabolism. In the acute phase of cerebrovascular disease, when the cerebral blood flow supply is drastically reduced, the OEF is elevated to take in more oxygen from the limited blood flow, a situation that is called "misery perfusion". In the subacute stage or later, cerebral blood flow is restored by vascular recanalization, but because cerebral tissue necrosis has already begun and oxygen is no longer needed, cerebral blood flow is supplied in excess of what is needed, which is called "luxury perfusion". In this phase, the OEF is lower than the normal value.

An up-and-coming French neurologist, Jean-Claude Baron, was at the forefront of describing the pathogenesis of poor perfusion in terms of elevated OEF, or "misery perfusion", and his idea became an important indicator for exploring therapies to restore cerebral blood flow (Baron et al. 1981). Here is a typical example of the time-course of a stroke patient evaluated with PET (Fig. 7.6). In the acute phase after the onset of the disease (8 h), cerebral blood flow is reduced, resulting in oxygen deprivation where brain tissue is still viable. Cerebral oxygen metabolism is maintained (Fig. 7.6; left). However, in the subacute phase, which lasts approximately 1 week (day 7), cerebral blood flow recovers due to cerebral vessel recanalization and/or collateral blood flow paths, but most neurons are already dead or necrotic, resulting in luxury perfusion where cerebral blood flow is overabundant (Fig. 7.6; middle). Furthermore, in the chronic phase (day 28), there is a matched perfusion phase where both oxygen supply by cerebral blood flow and oxygen demand are balanced at a level lower than normal (Fig. 7.6; right).

Thus, the oxygen extraction fraction (OEF) is approximately 0.4 in normal brain tissue but increases to 0.6–0.8 during the acute period of misery perfusion, and thereafter decreases to 0.2–0.3 during the subacute period of luxury perfusion. In the chronic phase, matched perfusion is achieved. The above balance of oxygen supply and demand in the pathological state is shown schematically in the acute, subacute, and chronic phases of ischemic cerebrovascular diseases (Fig. 7.7). Since PET with an oxygen-15-labeled tracer can evaluate this balance over time, the OEF can be used as an index to indirectly evaluate the "penumbra", or the boundary between healthy and necrotic tissue (Astrup et al. 1977).

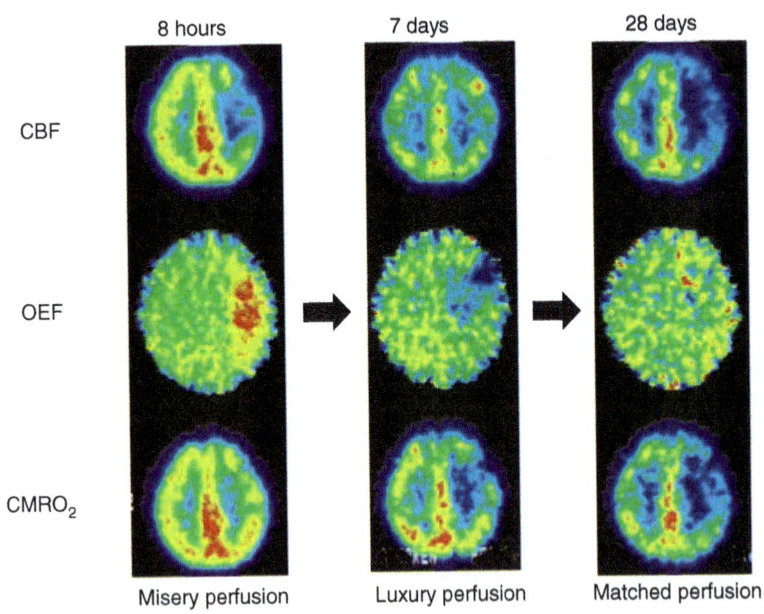

Fig. 7.6 Time course of cerebral circulation and oxygen metabolism after the onset of stroke: The cerebral blood flow (CBF), oxygen extraction fraction (OEF), and cerebral oxygen metabolism (CMRO2) were measured at 8 h, 7 days, and 28 days after the onset of middle cerebral artery (MCA) occlusion. The OEF in the occluded region turned from misery perfusion in the acute phase, to luxury perfusion in the subacute phase, and finally matched perfusion in the chronic phase

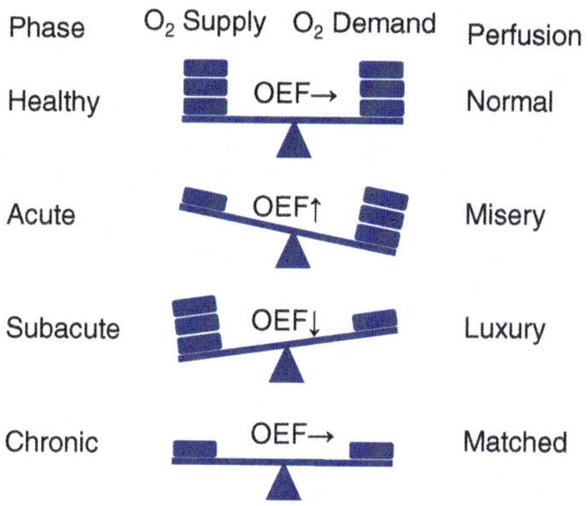

Fig. 7.7 Schematic illustrating how the OEF varies after a cerebrovascular incident: OEF is a parameter used to evaluate the balance between oxygen supply and demand. After the onset, the acute, subacute and chronic stages typically correspond to misery perfusion, luxury perfusion and matched perfusion, respectively

Fig. 7.8 Lucas Yamamoto (1928–2003)

Column: Lucas Yamamoto Introduced Young Japanese Neurosurgeons to the World

Lucas Yamamoto was a Japanese neurosurgeon (Fig. 7.8) who developed the PET system at the Montreal Neurological Institute (MNI) in the early 1970s and started clinical application. The author visited Yamamoto at the MNI several times, and he was welcomed warmly each time. Yamamoto implemented a method of measuring cerebral blood flow using krypton-77. The author visited Montreal and learned a lot about the PET system there from the researchers in MNI's laboratory. He was warmly welcomed because he had developed the Kanno-Lassen method, which calculates cerebral blood flow from the clearance curve measured via tomography of a RI inert gas. MNI developed the krypton-77 brain clearance method, in which krypton-77 condensed on the wall of a small glass tube is inhaled like a bolus, and the clearance curve is measured by the ring-type PET system to calculate cerebral blood flow.

7.3.2 Circulation Reserve and Metabolic Reserve

Let us consider the pathology of cerebral ischemia in terms of the remaining margin, or reserve capacity, of the cerebral circulation and metabolism. When the brain has a cerebrovascular deficit, the cerebral perfusion pressure (CPP) decreases distal (downstream) to the injured vessel. When this happens, an autoregulation mechanism is activated, causing cerebral blood vessels to dilate locally to maintain cerebral blood flow. Nevertheless, when the distal CPP reduces and reaches the lower limit of the cerebral autoregulation mechanism, the cerebral blood vessels cannot expand further (Fig. 7.9). That is, the ability to dilate blood vessels (circulation reserve) tends toward zero (depleted). If cerebral perfusion is further impaired in a situation where circulation reserve is depleted, CPP falls further, and cerebral blood flow begins to decline. When this happens, the OEF increases because the brain tissue takes in more oxygen from the decreased cerebral blood flow to maintain the required cerebral oxygen metabolism in brain tissue. In other words, the margin for oxygen uptake (metabolic reserve) begins to decline. As the CPP decreases further, the increase in OEF reaches its limit, and the metabolic reserve is depleted,

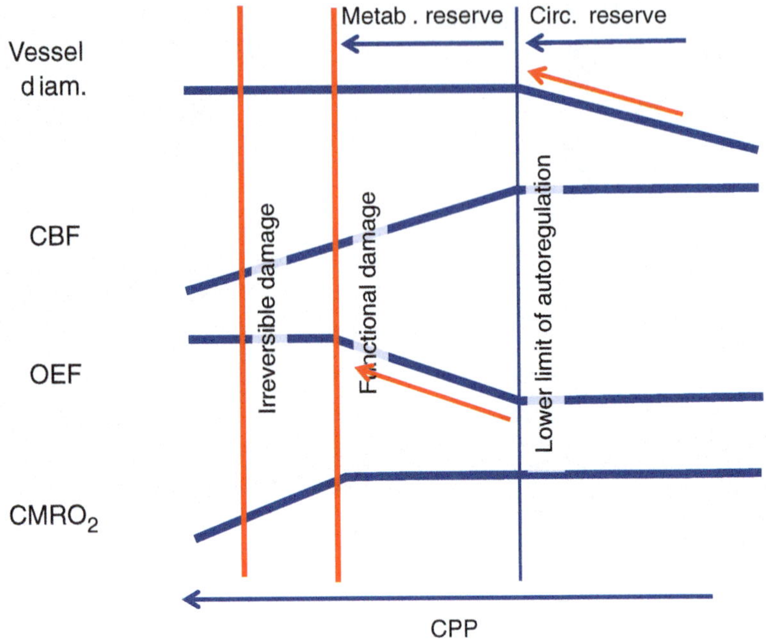

Fig. 7.9 Circulatory reserve and metabolic reserve: After vascular deficit, cerebral perfusion pressure (CPP) in distal areas gradually decreases (leftward). Within the autoregulation range cerebral blood flow (CBF) is maintained by dilation of vessel diameter (decrease in circulation reserve), but further decrease in CPP causes a decrease in CBF and elevation of OEF to maintain CMRO2 (decrease in metabolism reserve), resulting in functional damage. Further decrease in CPP results in irreversible damage (infarction)

resulting in brain dysfunction (functional failure). As CPP continues to decrease, neuronal membrane functions such as ion channels begin to fail, causing irreversible changes and tissue damage (cell death) that make it difficult for brain tissue to survive. In actual cerebrovascular disease, CPP is thought to be in a state where the various ischemic levels described above occur in patches depending on the local blood flow situation, so the regions where there is a depletion of circulation reserve or a depletion of metabolic reserve are intertwined.

In cerebrovascular disease, it is important to know the details of circulation reserve and metabolic reserve to understand the status of CPP. First, cerebral blood flow is measured in the resting state, and then cerebral blood flow under vasodilatation with acetazolamide (ACZ) or hypercapnia is measured. If there is no increase in cerebral blood flow with vasodilatation, it means that the cerebral blood vessels are already maximally dilated, and the circulation reserve is depleted. With such measurements it will be possible to assess improvement in regional CPP before and after cerebral revascularization surgery. It is also essential to evaluate regional CPP before cerebral thrombolysis and neurosurgical treatment (Fig. 7.10). Iodine-123-labeled IMP SPECT is widely used as a method to quantitatively evaluate regional CPP, as it enables repeatable measurement of cerebral blood flow even at facilities without PET (see Sect. 6.1.2).

It is also thought that areas of reduced circulation reserve and reduced capacity for cerebral vasodilation are already intermingled with areas of reduced metabolic reserve. In other words, there are areas where OEF is elevated and areas where the cerebral ischemic situation is critical due to decreased perfusion pressure. OEF is an index measured by PET, but by combining PET and SPECT data in various cases, it has become possible to estimate OEF, i.e., metabolic reserve, to some extent from SPECT circulation reserve and cerebral blood flow levels without direct PET OEF data. It is also possible to estimate the risk of bleeding in cerebral ischemic regions to some extent. In both PET and SPECT, it is important to correctly assess the circulatory and metabolic

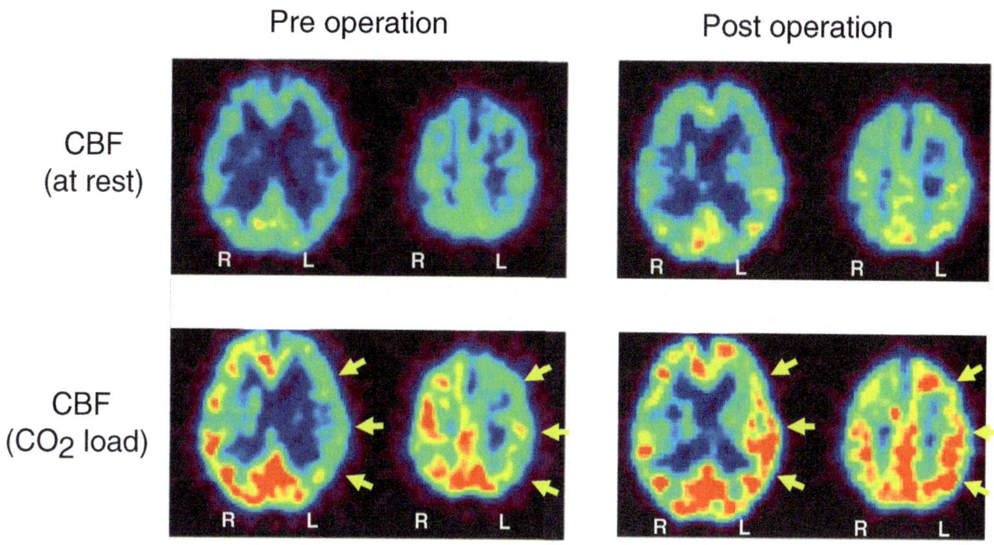

Fig. 7.10 Recovery of circulatory reserve after an operation to insert a stent into the left carotid artery: (Top row) At rest, cerebral blood flow is the same before and after the stent operation. (Bottom row) During CO_2 loading, the left hemisphere (arrowed side) did not dilate before the stent operation (left), but the dilation was the same as the contralateral side after the stent operation (right)

reserve in cerebrovascular diseases to estimate prognosis and determine the treatment plan.

7.3.3 Types of Cerebrovascular Occlusion

Of the three major types of cerebrovascular diseases, subarachnoid hemorrhage, intracerebral hemorrhage and cerebral infarction, the last is the most frequent. There are two types of cerebral infarction in terms of the way cerebral blood vessels are occluded. One is cerebral embolism, in which a relatively large thrombus formed in the intracardiac ventricle space or in the carotid artery, is released and embolizes the distal cerebral artery. In this case, blood flow distal (downstream) to the site of the embolus is suddenly interrupted. Cardiac diseases such as atrial fibrillation tend to cause plaque (blood clots) to form in the heart cavity, which can fly out and cause a cerebral embolism. This is called cardiogenic cerebral embolism. These cerebral embolisms, including plaques in the carotid artery, have a high probability of embolizing the main trunk of the cerebral artery, thus reducing cerebral blood flow over a wide area and causing serious cerebral ischemic injury. If treatment begins sufficiently early after embolization, recent advances in catheter technology have made it possible to safely remove the clot, thereby reducing the area of cerebral infarction.

Another type of cerebral infarction is cerebral thrombosis, in which the cerebral artery itself is narrowed or occluded due to the growth of atherothrombi in the cerebral artery. In cerebral thrombosis, also known as atherothrombosis, atheroma generally grows on a part of the vessel wall, and stenosis or occlusion progresses slowly, resulting in a gradual decrease in perfusion pressure on the distal (downstream) side. In response, collateral blood vessels, such as a pial artery-cerebral artery anastomosis, tend to develop. As a result, cerebral ischemic injury is relatively mild in many cases. However, if the vessel in which the cerebral thrombus originated is the main trunk of the cerebral artery, then, like the case for an embolism, a large area may be exposed to cerebral ischemia. In addition, since the circulation reserve is often lowered due to a decrease in regional CPP, it is necessary to measure the circulation reserve in advance and, if necessary, perform the treatment to improve blood circulation described in the next section.

Furthermore, cerebral infarction includes lacunar infarction, in which a narrow perforating branch artery called the lacunar vessel is narrowed and occluded due to atherothrombosis. In the case of lacunar infarction, the watershed area (see Sect. 3.3.2) is relatively small and may be asymptomatic without any obvious symptoms. In addition, chronic exposure to high blood pressure can lead to cerebral infarction in the white matter area surrounding the ventricles. If the white matter consisting of nerve fibers surrounding these ventricles is affected, cognitive function will deteriorate and often progress to so-called vascular dementia. Cerebral infarction caused by atherothrombosis of cerebral blood vessels is a typical geriatric disease and can be prevented by blood pressure control and dietary management to control lipids in the blood that cause atheroma.

7.3.4 Reperfusion Therapy for Cerebral Blood Flow

When the main cerebral artery is occluded, the severity of cerebral infarction is determined by the degree of cerebral blood flow reduction and the duration of exposure to ischemia. In cases of cerebral infarction, the extent of cerebral ischemia and circulation reserve should be promptly assessed. Based on the assessment, in serious cases cerebral blood flow reperfusion therapy is performed as soon as possible to restore cerebral blood flow to rescue brain tissue. Cerebral blood flow reperfusion treatment methods include surgical revascularization, medical thrombolysis, and more recently, vascular surgery for vasodilation and thrombectomy.

Surgical revascularization (so-called bypass surgery) includes STA-MCA, which anastomoses the extracranial shallow temporal artery (STA) to the intracranial middle cerebral artery (MCA), and EDAS, which places the temporal muscle on the brain surface for indirect revascularization in cases such as Moyamoya disease. These procedures are performed after confirming beforehand that cerebral blood flow in the anastomotic area is maintained to a certain extent and that brain tissue can be saved by bypass surgery. In addition, it is important to quantitatively evaluate the long-term changes in cerebral blood flow after surgery.

In terms of internal medicine, thrombolysis using tissue-plasminogen activator (t-PA) was started in the 2000s. Initially, to avoid the risk of complications of cerebral hemorrhage associated with reperfusion as described below, the application was limited to hyperacute cases within 3 h of the onset of cerebrovascular deficit (cerebral embolus or cerebral thrombus). However, the requirement of starting treatment within 3 h of onset is difficult to meet in a clinical protocol when the number of applicable cases is too small, so the time to start treatment was extended to 4.5 h from onset.

With the advancement of vascular catheters, the scope of application of vascular surgical treatment has expanded. Vascular catheterization is now being applied not only to the main cerebral arteries but also to small branching cerebral arteries. If necessary, it is used in combination with t-PA thrombolysis. In addition, thrombectomy, which removes fresh blood clots in blood vessels, has been developed, and under the slogan "Time is Brain," there is an urgent need to establish an emergency system to remove clots as soon as possible.

These therapies for cerebral blood flow reperfusion result in a complication called hyperperfusion syndrome, which is an excessive increase in cerebral blood flow in the ischemic region after the resumption. This is because high perfusion pressure blood flow is supplied to areas where autoregulation mechanisms are locally disrupted due to prolonged exposure to low perfusion, increasing the risk of cerebral hemorrhage in areas where BBB function is locally impaired. To prevent this, it is important to administer brain tissue protective agents in advance and to use tissue protection therapy in combination with reperfusion treatment. It is also essential to measure cerebral blood flow quantitatively beforehand since the level of low cerebral blood flow and the duration of exposure to low cerebral blood flow are used as a guide. These factors have been the driving force behind the development of quantitative cerebral blood flow measurement methods.

7.4 Cerebral Blood Flow During Problems on the Distal Side

In the last section, we mentioned the behavior of cerebral blood flow when the proximal (upstream) side of the brain is exposed to cerebrovascular deficits. Here, we will summarize the behavior of cerebral blood flow when the distal (downstream) side of brain tissue has a problem due to deterioration of neural function.

7.4.1 No-Reflow Phenomenon

This is a unique phenomenon that occurs after cerebral ischemia. It was discovered in an experimental rat model of stroke, in which some cerebral blood vessels in the rat are temporarily ligated to create a vascular occlusion model that cuts off blood flow to the distal region for a short time (30 min plus several tens of minutes). Cerebral blood flow is resumed after untying and releasing the cerebral vessels. This model reproduces the spontaneous recanalization phenomenon that often occurs in clinical practice and, since the upstream portion of the occluded vessel is open, the cerebral blood flow on the distal (downstream) side should reflow. However, when examined in detail most vessels have no reflow. This is called the no-reflow phenomenon (Ames 3rd et al. 1968). When looked at under a microscope, it was found that the vessels had not reopened at the capillary level. For many years, the mechanism for why no-reflow occurs at the capillary level was not understood, but recently, a report has been published that solves this mystery. At the capillary level, pericytes regulate blood flow through capillaries like smooth muscle, but when blood flow is temporarily blocked from the proximal side, the pericytes, which are vulnerable to ischemia, die while the capillaries remain closed so that they cannot reopen (Yemisci et al. 2009). However, the morphology and function of pericytes are still controversial, and it is unclear why they die prematurely in response to ischemic stress, so we need to continue to pay attention to the progress of research (Attwell et al. 2016).

7.4.2 Cerebral Blood Flow in Neurodegenerative Diseases

Neurodegenerative diseases are diseases in which specific groups of neurons in the brain or spinal cord (for example, neurons related to motor functions or cognitive functions) are gradually damaged and lost. Most are congenital diseases caused by specific genes, but only a small percentage of these diseases have known cures even if the cause at the genetic level is understood. The neurons and areas that drop out vary depending on the disease, but the symptoms can be broadly categorized into diseases that prevent smooth movement (Parkinson's disease, Parkinson's syndrome, etc.), diseases that make it difficult to maintain body balance (spinocerebellar degeneration, progressive supranuclear palsy, etc.), diseases that cause muscle weakness (amyotrophic lateral sclerosis: ALS), and cognitive decline (Alzheimer's disease, dementia with Lewy bodies, cortical basal ganglia degeneration, etc.). While each of these neurodegenerative diseases has its own pathogenesis, aging itself is thought to be a major cause, as the diseases tend to occur more frequently in elderly individuals. As aging progresses, abnormal proteins accumulate in the brain's nerves, causing a decline in the function of neurons, which eventually leads to glial and neuronal loss, which in turn impairs the neural network and leads to dementia. Since the areas affected by degenerative diseases are often distinctive, it is possible to diagnose degenerative disease from the distribution pattern of cerebral blood flow measured during the early stage of disease onset. The diagnosis of dementia with cerebral blood flow is a useful tool for the differential diagnosis of dementia, although it is, so to speak, an observation of burned-out neurons.

7.4.3 Cerebral Blood Flow in Dementia with Abnormal Protein Accumulation

In nuclear medicine, there is an ongoing attempt to image the expression of abnormal proteins before the onset of dementia. Beta-amyloid has been found to accumulate in the brains of Alzheimer's patients long before the onset of the disease (Jack et al. 2010). There was fierce competition to develop RI tracers for imaging amyloid, and C-11-labeled PIB was developed at the University of Pittsburgh in the early 2000s (Klunk et al. 2004) (Fig. 7.11). Subsequently, several tracers for imaging beta-amyloid were developed, including F-18-labeled florbetapir and F-18-labeled flutemetamol, which can both be used at facilities without a cyclotron because of the long half-life of F-18, and they are now commercially available. Early diagnosis by observation of abnormal protein accumulation was also attempted. However, later studies showed that amyloid accumulation was not consistent with dementia symptoms and that beta-amyloid imaging was consequently only ancillary information.

The next target was tau protein imaging, which is said to appropriately reflect dementia severity and disease type (Fig. 7.12) (Maruyama et al. 2013). F-18-labeled THK5351 from Tohoku University, C-11-labeled PBB3 from NIRS, and F-18-labeled AV1451 from an American pharmaceutical company have been developed and were followed by the development of next-generation tracers with further improvements. Clinical trials to eliminate abnormal protein accumulations have begun, and confirmation of the clinical therapeutic efficacy and evaluation of therapeutic agents using these imaging techniques are proceeding in parallel. In addition, the development of tracers for imaging abnormal proteins accumulated in neurons, such as alpha-synuclein and TDP-43 (TAR-DNA-binding protein of 43 kDa), has begun. The relationship between the distribution of abnormal proteins in the brain and various types of dementia and neuropsychiatric disorders has also been studied. The abnormal proteins that accumulate in the various types of dementia are thought to be washed out by cerebral blood flow. However, some studies have reported that abnormal protein accumulation is increased by abnormalities in the cerebral circulatory and metabolic environment of the brain tissue (Nedergaard 2013) (see Sect. 2.2). The accumulation and washing out of abnormal proteins in the brain has become an important theme in elucidating the biochemical mechanisms of dementia.

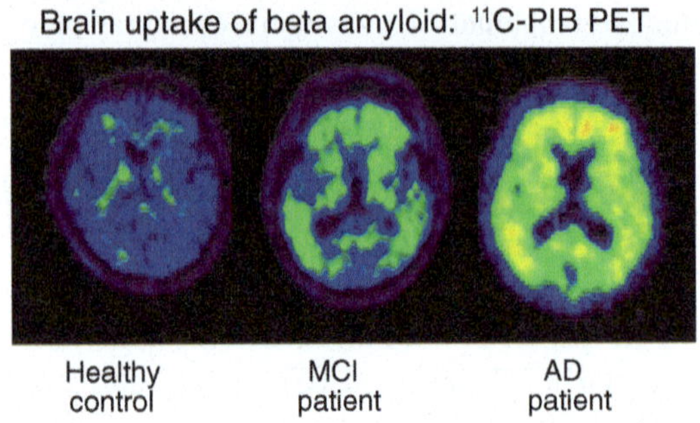

Fig. 7.11 Beta amyloid uptake measured using PET and PiB: PiB-PET images in a healthy control (left), MCI (mild cognitive impairment) patient (middle), and AD (Alzheimer disease) patient (right). Beta amyloid uptake increases in order from the control to the MCI patient and then to the AD patient

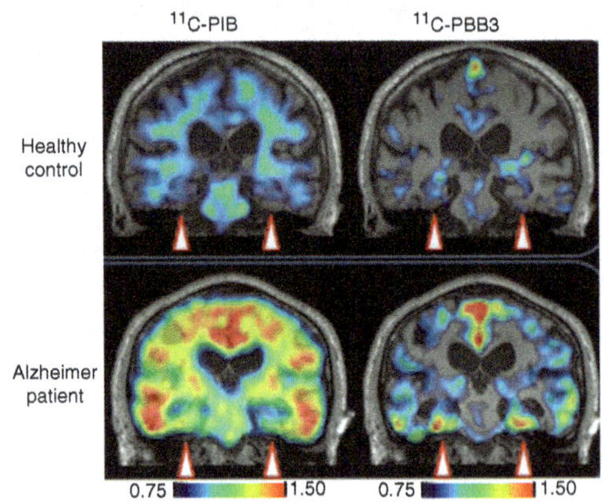

Fig. 7.12 Beta-amyloid uptake of C11-PIB (left), and tau protein uptake of C11-PBB3 (right): Comparison between a healthy adult (top row) and an Alzheimer patient (bottom row). (Maruyama M et al., Neuron 2013)

7.4.4 Cerebral Blood Flow in Vascular Dementia

Some types of dementia are thought to be caused by decreased cerebral blood flow in brain microvessels. These forms of dementia can be caused by a small infarction called "asymptomatic cerebral infarction" without any subjective symptoms, or they can progress stepwise due to extensive cerebrovascular disorders such as cerebral infarction or cerebral hemorrhage. In addition, insufficient cerebral blood flow in microvessels, such as arterioles, gradually progresses, resulting in blood flow impairment in the cerebrovascular network of the whole brain. In many cases, the disease is diagnosed as mixed vascular dementia combined with Alzheimer's disease as well, and a variety of cognitive symptoms occur. Vascular dementia can be classified into cortical, thalamic, microvascular, hypoperfusion, and cerebral hemorrhagic depending on the site of decreased blood flow in the cerebral microvessels, and the onset of cognitive impairment differs. In addition to cognitive impairment, vascular dementia often causes complications with physical neurological symptoms such as numbness, paralysis, and gait disturbance and causes a patchy pattern of symptoms, such as "memory impairment with preserved judgment", depending on the area of cerebrovascular damage. It is also well known that so-called white-matter lesions, in which the white matter around the lateral ventricles becomes hypoperfused, are linked to dementia. Elucidation of the relationship between cerebrovascular disease and cognitive impairment will be an important issue in the future.

References

Ames A 3rd, Wright RL, et al. Cerebral ischemia. II. The no-reflow phenomenon. Am J Pathol. 1968;52(2):437–53.

Astrup J, Symon L, et al. Cortical evoked potential and extracellular K+ and H+ at critical levels of brain ischemia. Stroke. 1977;8(1):51–7.

Attwell D, Mishra A, et al. What is a pericyte? J Cereb Blood Flow Metab. 2016;36(2):451–5. Review

Baron JC, Bousser MG, et al. Reversal of focal "misery-perfusion syndrome" by extraintracranial arterial bypass in hemodynamic cerebral ischemia. A case study with ^{15}O positron emission tomography. Stroke. 1981;12(4):454–9.

Gotoh F, et al. Dual control of cerebral circulation: Separate sites of action in vascular tree in autoregulation and chemical control. In: Langfitt TW, et al., editors. Cerebral circulation and metabolism. New York: SpringerVerlag; 1975. p. 43–5.

Hotta H, Masamoto K, et al. Layer-specific dilation of penetrating arteries induced by stimulation of the nucleus basalis of Meynert in the mouse frontal cortex. J Cereb Blood Flow Metab. 2013;33(9):1440–7.

Jack CR Jr, Knopman DS, et al. Hypothetical model of dynamic biomarkers of the Alzheimer's pathological cascade. Lancet Neurol. 2010;9(1):119–28.

Klunk WE, Engler H, et al. Imaging brain amyloid in Alzheimer's disease with Pittsburgh Compound-B. Ann Neurol. 2004;55(3):306–19.

Lassen NA. Brain extracellular pH: the main factor controlling cerebral blood flow. Scand J Clin Lab Invest. 1968;22(4):247–51.

Lassen NA. Control of cerebral circulation in health and disease. Circ Res. 1974;34(6):749–60. Review

Maruyama M, Shimada H, et al. Imaging of tau pathology in a tauopathy mouse model and in Alzheimer patients compared to normal controls. Neuron. 2013;79(6):1094–108.

Nedergaard M. Neuroscience. Garbage truck of the brain. Science. 2013;340(6140):1529–30.

Snyder GK. Erythrocyte evolution: the significance of the Fåhraeus-Lindqvist phenomenon. Respir Physiol. 1973;19(3):271–8.

Stromberg DD, Fox JR. Pressures in the pial arterial microcirculation of the cat during changes in systemic arterial blood pressure. Circ Res. 1972;31(2):229–39.

Yemisci M, Gursoy-Ozdemir Y, et al. Pericyte contraction induced by oxidative-nitrative stress impairs capillary reflow despite successful opening of an occluded cerebral artery. Nat Med. 2009;15(9):1031–7.

8 Cerebral Blood Flow Mapping the Brain

Abstract

Cerebral blood flow reflects the neural activity of the brain. This chapter discusses the mechanism by which neural activity is translated into cerebral blood flow. Initially, it was naively believed that cerebral blood flow increased to support the brain during activity from a feedback theory point of view. However, after many experiments and much discussion of cerebral circulation physiology, it was eventually realized that feedback theory does not fit to the real situation. A feed forward mechanism based on vascular mediators was shown to be more likely. In addition, the discovery of BOLD contrast with MRI opened a new platform for neuroscience research and there was a rapid increase in the neuroscientist population. Using BOLD contrast has widened the field of brain function mapping. The research paradigm has changed from finding activity foci in the brain with PET, to mapping networks of regions that are somehow connected during brain activity even during the resting condition. This chapter will discuss the physiological mechanism by which cerebral blood flow reflects brain activity and explores the mystery of why cerebral blood flow maps brain function in so much detail.

Keywords

Functional mapping · Aerobic glycolysis · Anaerobic glycolysis · BOLD contrast · Oxygenation · Deoxygenation · Neurovascular coupling · Penfield homunculus

8.1 Cerebral Blood Flow Linked to Neural Activity

When the brain is active, cerebral blood flow around the active area increases. There is a one-to-one correspondence between the area of brain activity and the area where cerebral blood flow increases. This phenomenon is called neurovascular coupling (NVC). First, let us explore the physiology behind it.

8.1.1 Feedback Theory

When Lassen first succeeded in capturing human brain activity as a map (see Sect. 2.5; Fig. 2.4), he proposed a simple explanation for the mechanism by which cerebral blood flow increases during brain activation. Namely, he explained that when we perform an activation

I. Kanno, *What Cerebral Blood Flow Tells Us*, https://doi.org/10.1007/978-981-97-6382-5_8

task, neurons use energy to carry out the task, and oxidative respiration of glucose generates CO_2, which locally dilates the surrounding cerebral blood vessels and increases cerebral blood flow to the necessary extent. Such a feedback mechanism was naturally accepted at the time. Soon after, Phelps and colleagues at the University of California, Los Angeles (UCLA) measured brain glucose metabolism using PET and FDG and found that brain glucose metabolism increased locally during functional activation for various stimulus tasks (Phelps and Mazziotta 1985). Brain activation increases cerebral glucose metabolism, which in turn increases cerebral blood flow. These findings were consistent with Lassen's explanation. Thus, it was long believed that brain activity increases the consumption of metabolic substrates in the brain and subsequently cerebral blood flow increases to replenish them, and no one doubted this mechanism (Fig. 8.1). Such a classical feedback mechanism was reasonable and widely accepted because it was natural to think of the balance between energy supply and energy consumption associated with brain activity. At that time, however, it was unclear how precisely brain activity and changes in cerebral blood flow corresponded spatially and temporally, and it was not until in vivo microscopic measurements were realized that the detailed relationship became clear (see Sect. 8.3).

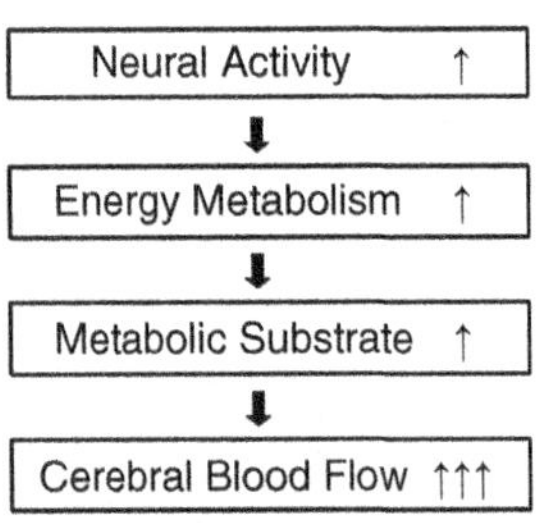

Fig. 8.1 Classical hypothesis for neurovascular coupling: Brain stimulation increases neural activity, causing an increase in energy metabolism, elevating demand for metabolic substrate, and finally resulting in an increase in cerebral blood flow. Thus, this hypothesis is based on a feedback system

8.1.2 Failure of the Feedback Theory

To verify the feedback mechanism, the author's group performed cerebral blood flow measurements that combined a simple neuronal stimulus task with various physiological loadings. First, visual stimulation was performed while inhaling CO_2 to induce hypercapnia where cerebral blood flow is globally increased. It is well known that hypercapnia increases the cerebral blood flow without affecting brain metabolism. Therefore, it was predicted that the increase in cerebral blood flow during visual stimulation would be reduced because the cerebral blood flow had already been increased by the hypercapnia condition. However, the increase in cerebral blood flow was greater than the increase in cerebral blood flow during normocapnia (normal breathing) (Fig. 8.2) (Shimosegawa et al. 1995). Visual stimulation was also performed under hyperoxia ($PaO_2 = 500$ mmHg) with high O_2 concentration (100%) inhalation to increase the oxygen content in the blood. At the time, it was predicted that the increase in cerebral blood flow due to brain activation would decrease because the blood oxygen content had already increased compared to that during normal breathing. In fact, contrary to the prediction, the increase in cerebral blood flow due to brain activation under hyperoxia increased more than that during normoxia (normal breathing) (Fig. 8.3) (Kanno unpublished data). Similar experiments were also performed in rats using electrical stimulation of the hindpaw, and LDF and confirmed similar changes to the cerebral blood flow as for humans. It was also directly verified using local field potential (LFP) measurements that neural activity in the cerebral cortex was unaffected by the physiological respiratory loadings (Matsuura et al. 2001). Thus, in both humans and animals, the amount that the cerebral blood flow increased was unaffected by the excessive supply of O_2. This suggests that the increase in cerebral blood flow due to brain activation is not solely for the purpose of supplying O_2 to brain tissue.

Prior to these experiments, Fox and Raichle reported a striking result from PET measure-

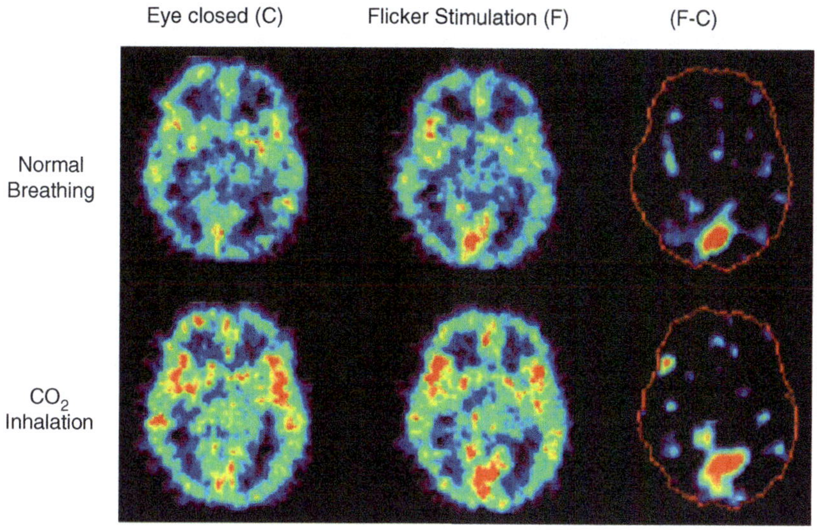

Fig. 8.2 Magnitude of cerebral blood flow response to stimulation depends on the baseline level: Cerebral blood flow in the visual cortex with the eyes closed (C), viewing a flicker stimulus (F), and the subtraction (F-C). The response (F-C) was higher during CO_2 inhalation (bottom row) than during normal breathing (upper row). (Modified from Shimosegawa E et al. J Cereb Blood Flow Metab 1995)

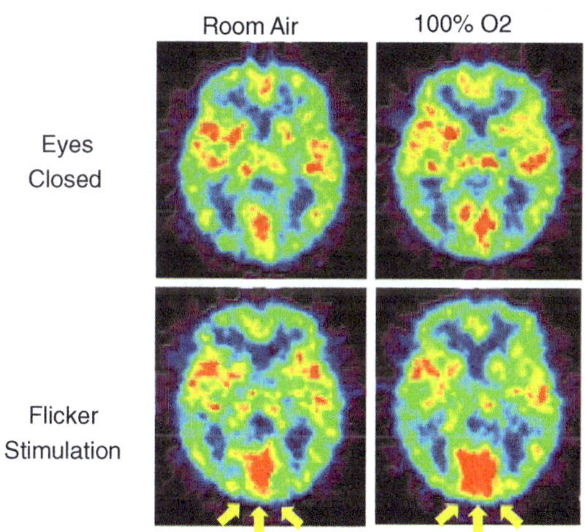

Fig. 8.3 Cerebral blood flow response during 100% oxygen inhalation: Contrary to the expectation that the cerebral blood flow response to flicker stimulation would have been suppressed during 100% oxygen inhalation, the response was not reduced, but rather enhanced (right) compared to that observed during inhalation of room air (left). (Kanno I et al., unpublished data)

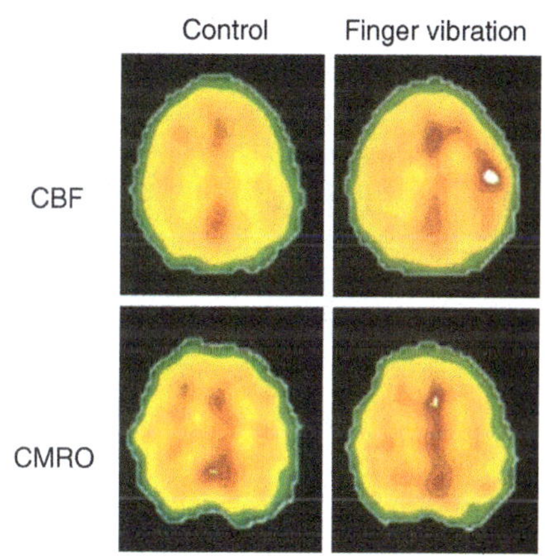

Fig. 8.4 Cerebral blood flow (CBF) and cerebral oxygen consumption (CMRO2) in response to finger vibration: CBF was clearly increased in the motor cortex area due to the stimulation, but CMRO2 did not increase. (Fox PT, Raichle ME, Proc Natl Acad Sci USA 1986)

ments (Fox and Raichle 1986). When hand vibration stimulation was applied during PET, cerebral blood flow increased by 20–30% in the somatosensory cortex, but cerebral oxygen metabolism only increased by 5% (Fig. 8.4). This was a striking overthrow of the classical feedback-based theory that cerebral blood flow augments to increase metabolic substrates in brain-activated areas. Fox et al. then measured the cerebral blood flow, the cerebral oxygen metabolic rate and the cerebral glucose metabolic rate under visual stimulation to confirm their findings. They found that cerebral blood flow and cerebral glucose

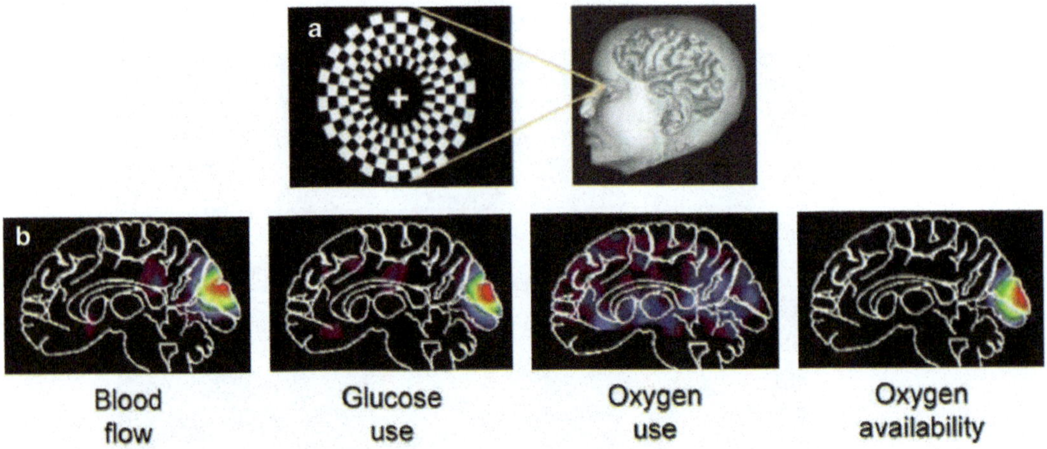

Fig. 8.5 Viewpoint fixation versus checkerboard stimulation: Visual stimulation strongly elevated blood flow and glucose metabolism but there was little increase in oxygen metabolism. Oxygen availability is calculated by multiplying cerebral blood flow by arterial oxygen content. (Modified from Raichle ME, Mintun MA, Ann Rev. Neurosci 2006)

metabolism increased at the same rate, but cerebral oxygen metabolism increased only slightly (Fox et al. 1988). This was a surprising result that overturned the existing theory of cerebral physiology, which had been established in the resting-state since the time of Kety, where cerebral oxygen metabolism, cerebral blood flow, and cerebral glucose metabolism are coupled. The brain metabolic rates of glucose and oxygen, which were thought to increase in parallel as energy substrates, dissociated. Fox et al. interpreted their results to mean that anaerobic glycolysis of glucose, which provides energy even in the absence of oxygen, proceeded during the brain activation task (Fig. 8.5) (Raichle 1998).

The PET data reported by Fox and Raichle in 1986 overturned the classical feedback theory that increased cerebral blood flow was due to an increased need for energy substrates. Sokoloff, a giant in brain physiology, had previously confirmed with autoradiography measurement of animals that regional glucose metabolism (using carbon-14-labeled DG) and regional cerebral blood flow (using carbon-14-labeled IAP) were coupled. He found it difficult to accept Fox and Raichle's result that cerebral blood flow and cerebral oxygen metabolism were uncoupled, and he argued until he was red-faced that the oxygen-15 measurement model for PET was wrong. However, it should be noted that Sokoloff's experiment measured cerebral glucose metabolism during the steady-state, while Fox and Raichle's study measured cerebral glucose metabolism during the activation state of the brain. At the time, measurements of cerebral blood flow, cerebral oxygen metabolism, and cerebral glucose metabolism in the activated state and in the resting steady-state were discussed as if the two states were on the same platform so they may have been confused (Frackowiak et al. 2001). Since then, neurophysiologists have been busy establishing a new theory for the relationship between cerebral blood flow and cerebral oxygen metabolism.

Column: Raichle Opens Pandora's Box

In 1990, a Benzon Symposium organized by Lassen, called Brain Work II, was held in Copenhagen. Following the first Brain Work meeting held 15 years earlier in 1975, this was the second Benzon Symposium to discuss the same topic. In the general discussion at the end of Brain Work II, Raichle quietly said this about neurovascular coupling: "It is amazing that basic research has developed so much in the last 15 years, and yet the mechanism has not been elucidated.

If we hold such a symposium in the next 15 years, we do not want to repeat that it is unsolved" (Brain Work II 1990). Fifteen years later, in 2005, there were still no concrete findings on neurovascular coupling. This may have been the driving force behind Raichle's idea of the default mode network (DMN), which will be discussed in the next section, and the question he had been asking himself for a long time, "What is the resting state?" When Raichle was invited to Japan to give a lecture at RIKEN a few years ago, he spoke passionately about his recent work on default mode networks.

8.1.3 Aerobic Glycolysis During Neural Activation

As well as the regulation of cerebral blood flow during neural activation, the phenomenon that the increase in cerebral oxygen metabolism is far less than the increase in cerebral blood flow has been a headache for neurophysiologists since Fox and Raichle's 1986 report. In other words, the increases in cerebral blood flow and cerebral glucose metabolism are much higher than the increase in cerebral oxygen metabolism, which is a mismatch between cerebral oxygen metabolism and cerebral glucose metabolism. In response to that finding, Raichle and colleagues proposed a new interpretation of glycolysis. In the absence of O_2, glucose normally undergoes anaerobic glycolysis and is broken down into pyruvate to produce three adenosine triphosphates (ATP). If there is sufficient O_2, the pyruvate is broken down into CO_2 and water through the electron transfer system in the tricarboxylic acid (TCA) cycle, and finally, 32 ATPs are produced. The ATP produced at this stage is used to form synapses between neurons and is thought to contribute to the formation of novel neural networks (Raichle and Mintun 2006).

In a meta-analysis of previous brain metabolism studies, Goyal and Raichle found that cerebral blood flow, cerebral glucose metabolism, and cerebral oxygen metabolism change with age. In comparison to adult levels, cerebral glucose metabolism and cerebral blood flow are maximal at birth and decline with age thereafter. On the other hand, cerebral oxygen metabolism is closer to the adult level at birth before gradually declining with age (Fig. 8.6)

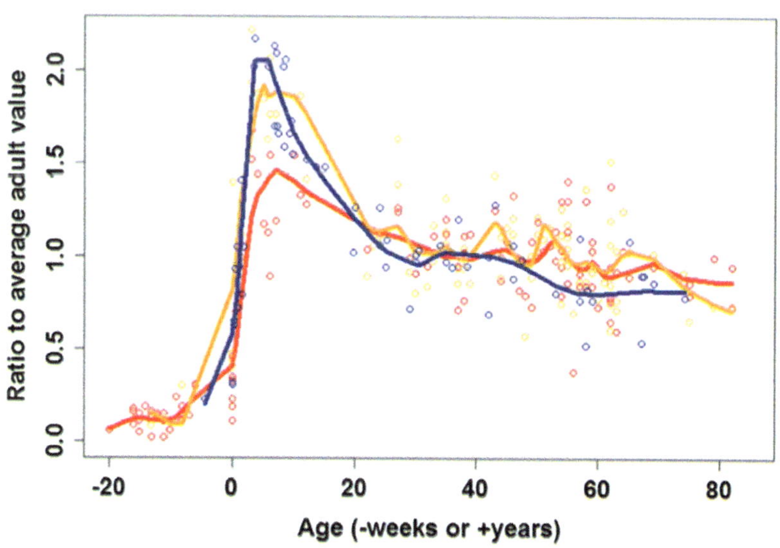

Fig. 8.6 Meta-analysis showing the relative changes in cerebral glucose metabolism (blue), cerebral oxygen metabolism (red), and cerebral blood flow (yellow), for the fetal stage, after birth, and through to old age. (Goyal MS et al. Cell Metab 2014)

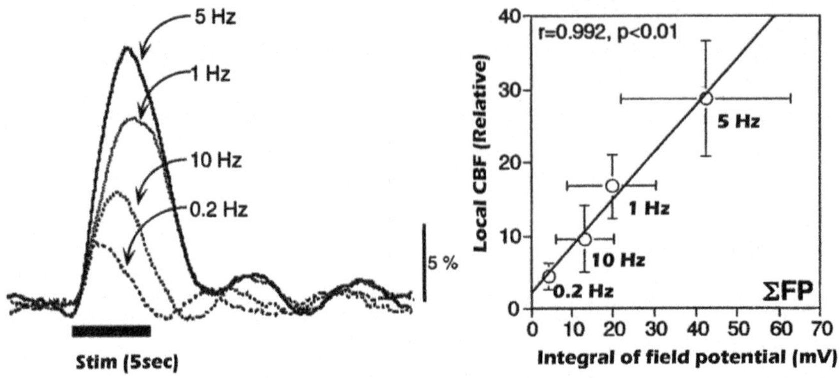

Fig. 8.7 Relationship between cerebral blood flow (CBF) and local EEG during hind-paw stimulation in rat: LDF response curve for hind-paw stimulation at various frequencies (left), and comparison of the integral of the CBF with the integral of the local field potential (FP) (right). (Modified from Matsuura T and Kanno I Neurosci Res 2001b)

(Goyal et al. 2014). This is consistent with the discrepancy between cerebral oxygen metabolism and cerebral glucose metabolism, i.e., anaerobic glycolysis is maximal at birth and synaptogenesis is maximal in early childhood, suggesting that anaerobic glycolysis corresponds to the synaptogenesis process during brain development.

8.1.4 Feed-Forward Theory Versus Feedback Theory

Since the results of Mosso and Roy & Sherrington near the end of the nineteenth century, many researchers have attempted to elucidate the mechanism regulating the cerebrovascular response during brain activation. However, it is now widely recognized that the increase in cerebral blood flow is not a feedback mechanism to increase nutrient supply. Rather, it is thought that vasoactive substances (mediators) associated with neurosynaptic activity, such as K^+, NO and prostanoids, are released by neurons during activation to regulate blood vessel response (Attwell and Iadecola 2002; Attwell et al. 2010). The release can be interpreted as turning on a faucet at the time of brain activation to increase cerebral blood flow. In other words, the release seems to dilate cerebral blood vessels in a feed-forward manner.

On the other hand, it has been macroscopically confirmed that the relationship between the intensity of regional brain activity and the fractional increase in regional cerebral blood flow is linear (Zheng et al. 2010). In fact, the analytical model of brain activity measurement is built on feedback theory, i.e., the linearity of the increase in cerebral blood flow in response to neural activity. In animal experiments, the author's group confirmed that the increase in cerebral blood flow in response to a simple stimulus is linearly proportional to the integral of the local field potential during neural activity (Fig. 8.7) (Matsuura and Kanno 2001). Animal experiments have also confirmed a linear relationship between the integral of the increase in cerebral blood flow during various blood flow states from vasodilation to contraction. However, it is not clear how and when the faucet is turned off to reduce cerebral blood flow to baseline after activation ceases. It is necessary to build a reliable model of how the NVU works. It will take some time to figure this out (Iadecola 2017).

8.2 BOLD Contrast: A New Paradigm

The discrepancy between cerebral blood flow and cerebral oxygen metabolism discovered by Fox and Raichle in the 1980s was only a prelude to

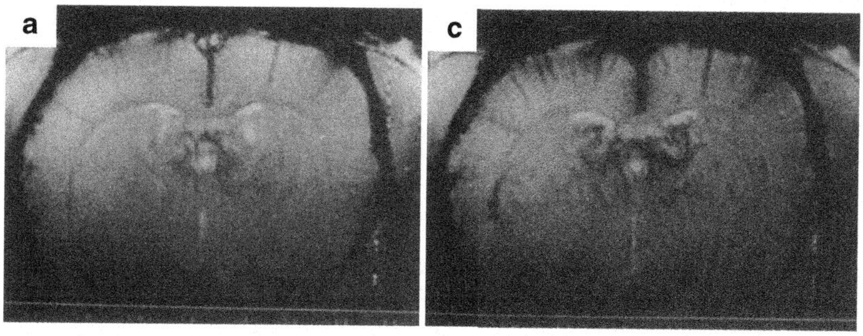

Fig. 8.8 Discovery of the blood oxygenation level dependent (BOLD) contrast signal: During MRI imaging of a coronal section of a rat (left), radial black lines appeared (right). The lines corresponded to deoxygenated hemoglobin after the death of the rat. This effect was named blood oxygenation level dependent (BOLD) contrast. (Ogawa S et al. Proc Natl Acad USA 1990)

the discovery by Ogawa and his colleagues in the 1990s of blood oxygen level-dependent (BOLD) contrast with MRI.

8.2.1 Discovery of the BOLD Contrast

Ogawa and colleagues discovered the BOLD contrast accidentally. During a routine MRI experiment on a rat brain, they observed radial black lines in the cerebral cortex that were not normally visible on coronal images of the brain (Fig. 8.8). It was later verified that the rats were dead at the time the images were acquired. Ogawa had measured hemoglobin with MRI many times, so he knew immediately what the black lines meant. The black lines were along the cerebral blood vessels, indicating that the hemoglobin in the cerebral blood vessels had changed to a paramagnetic substance called deoxygenated hemoglobin.

Ogawa and colleagues predicted that if they could use this phenomenon while performing neural activation, they would be able to see a signal reflecting increased deoxygenated hemoglobin in the veins during the increased cerebral oxygen metabolism associated with brain activation. They measured the MRI signal while applying electrical stimulation to the upper limbs of rats, but the results were not as they predicted. Based on the 1986 report by Fox and Raichle presenting the mismatch between cerebral blood flow and cerebral oxygen metabolism, they postulated that oxygenated hemoglobin in venous blood increases during brain activation. Ogawa and colleagues were convinced that MRI could measure neural activation in the brain using the BOLD contrast. BOLD is a real-time measurement of changes in magnetic susceptibility, as opposed to PET, which measures cerebral blood flow averaged over a couple of minutes as the tracer, oxygen-15-labeled water, passes through the brain. BOLD signal measurement time is far shorter than PET, less than a few seconds. After the discovery of Ogawa's new contrast, there was a robust discussion in the MRI research community about the relationship between the observed signal changes and the hemodynamic response. It was not immediately clear how factors such as changes in the composition of oxygenated and deoxygenated hemoglobin in cerebral arterial and venous blood and changes in cerebral blood volume would combine to form the BOLD signal during brain activation. It has now been established that changes in BOLD signal intensity reflect changes in the magnetic field environment around the vessels due to changes in hemoglobin oxygenation in the blood (Fig. 8.9). Ogawa and colleagues named the contrast they discovered the blood oxygen level dependent (BOLD) contrast (Ogawa et al. 1990), but it took some time and various experiments before everyone was convinced that they were right.

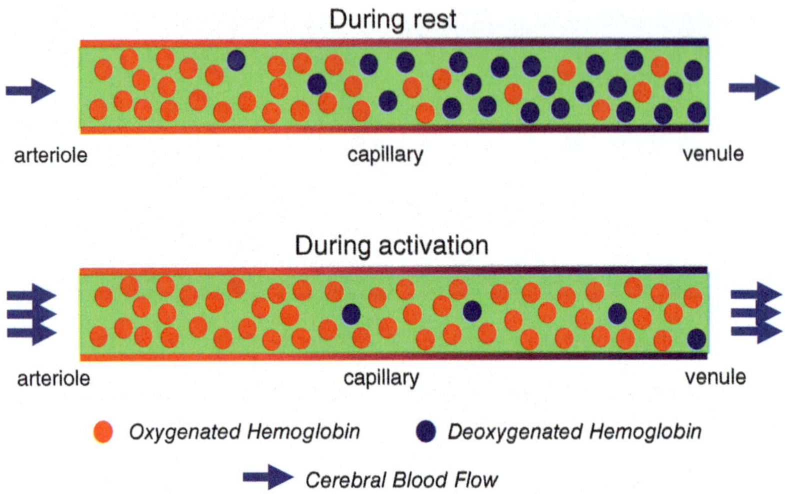

Fig. 8.9 Schematic illustrating the oxygenation and deoxygenation of capillary blood during rest (top) and activation (bottom): During rest the ratio of oxygenated hemoglobin (red) to deoxygenated hemoglobin (blue) is steady (top), but when the brain region is activated cerebral blood flow will increase and the ratio of oxygenated hemoglobin (red) to deoxygenated hemoglobin (blue) is increased (bottom)

Column: Ogawa, Who Opened the New Window

In the late 1980s, many PET facilities started to use cerebral blood flow measurement to map brain function. At the same time, MRI researchers were enthusiastically investigating the idea that, since cerebral blood flow could be measured using angiography, brain function mapping with MRI may also be possible. I heard of a Japanese researcher in the U.S.A who had used MRI to observe brain activation signals different from those of angiography. It was Seiji Ogawa and he had returned to Japan. I met Ogawa at the Japanese Society of Cerebral Blood Flow and Metabolism meeting in Tokyo and invited him to the Akita Cerebrovascular Research Center (Akita Brain Institute) to give a talk about his findings. Ogawa carefully explained his experiments, telling us how he was able to rule out changes in blood vessel diameter (i.e., cerebral blood volume) and changes in cerebral blood flow as being the source of the new contrast. He had persisted until he was convinced that the MRI signal source was reflecting an increase in oxygenated hemoglobin. As I was not familiar with MRI, I could not understand the physical details, but I understood that Ogawa was struggling to interpret the hemodynamics associated with brain activation. I was fortunate to meet Ogawa in the early days of the BOLD signal. His work inspired me to make understanding the mechanism of neurovascular coupling (NVC) linked to brain activation one of the goals of my research.

On the other hand, many neuroscientists did not at first realize the importance of the BOLD contrast report by Ogawa and colleagues. However, researchers who had previously investigated brain activation using PET easily understood that the BOLD signal was a signal corresponding to changes in cerebral blood flow during brain activity. Using PET, the author's group had already measured the hemodynamics of cerebral blood flow and knew that regional cerebral blood flow increases by 20–30% during brain activation and that the volume of the cerebral blood vessels (regional cerebral blood volume) increases by approximately one tenth (2–3%) of the cerebral blood flow. It was a good opportunity to discuss the mechanism of BOLD contrast with Ogawa and other MRI researchers based on the hemodynamics of brain blood vessels.

Following the success story of Ogawa's discovery, BOLD contrast imaging has become an indispensable tool for brain science research. However, the subsequent expansion and improvement of functional MRI (fMRI) based on BOLD signal far exceeded the predictions of the author and others in the PET field. Unlike PET, which requires a large cyclotron and many staff, fMRI with the BOLD contrast is a completely noninvasive method of investigating cerebral hemodynamics and oxygen metabolism, requiring only a clinical MRI system and a few staff members. The fact that fMRI is a completely noninvasive method with no radiation hazard means that it is superior to PET and has led to the explosive spread of fMRI.

8.2.2 The Physiology of BOLD Contrast

The essence of BOLD contrast during brain activation is an excessive increase in cerebral blood flow relative to cerebral oxygen metabolism that is accompanied by a decrease in the oxygen extraction fraction (OEF). In other words, the increase in cerebral blood flow associated with neural activity results in an increase in the concentration of oxygenated hemoglobin. During presentations Raichle frequently used a slide explaining the BOLD contrast. He showed neuronal activity on the X-axis, oxygen supply (= blood flow × arterial blood oxygen content) on the left Y-axis, and oxygen consumption on the right Y-axis, showing the discrepancy between the two Y-axes (Fig. 8.10) (Raichle 1998). If the quantities plotted on the Y-axes were in balance with the neuronal activity, the slope would be 45°. Raichle's diagram did indeed show that the oxygen supply (left Y-axis) was at 45° because the rate of increase is proportional to neural activity. However, the oxygen consumption (right Y-axis) had a slope smaller than 45° because its rate of increase is lower than that of neural activity, meaning a lowering in OEF. When at rest (= center of the diagram), blood flow and oxygen consumption are in balance, so both Y-axis quantities pass through there. Using such a diagram, Raichle was able to answer many of the questions he had been asking since the time when PET was the only tool used to measure human brain activity. For example, even in a state of little neuronal activity where there is no obvious brain activation, OEF decreases, and the concentration of oxygenated hemoglobin increases as the source

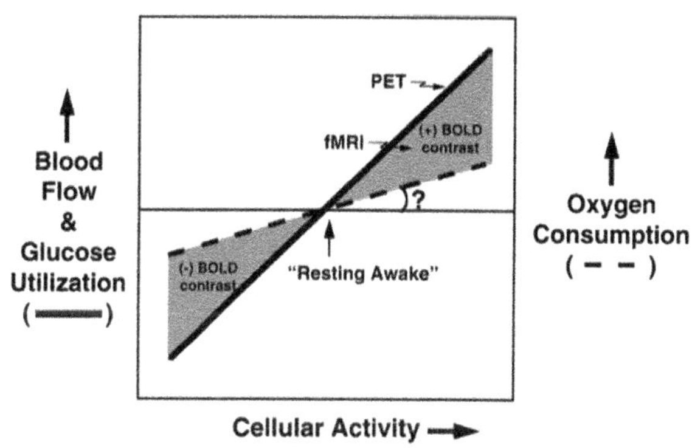

Fig. 8.10 Relationship between blood flow and oxygen metabolism: The slopes of the blood flow (left vertical axis) and the oxygen metabolism (right vertical axis) when plotted against neuronal activity (horizontal axis) are different. The shaded area corresponds to regions where the BOLD contrast occurs. (Modified from Raichle ME Proc Natl Acad Sci USA 1998)

of the BOLD signal. Thus, the BOLD signal increases slightly even when the brain is almost at rest. This led to the new discovery of resting-state fMRI (rsfMRI) (see Sect. 8.4), where it is thought that activity is continuously occurring in networks throughout the whole-brain.

Column: fMRI Experiments Were Always at Midnight

In the mid 1990s, the author tried to compare the BOLD response with the changes in cerebral blood flow measured with PET. The fMRI experiments were always performed at night after clinical studies were finished for the day because the high performance 1.5T MRI scanner with fast imaging capability (EPI = Echo Planar Imaging) was busy with patients until evening. We started preparations in the evening and finished the measurements before midnight, but the analysis and related work were not completed until just before the first clinical examination at 8:00 am the next morning. Using the same tasks that had been previously used in PET brain research, we were preoccupied with comparing the relative change in PET and BOLD signal, as well as verifying that the anatomical locations agreed between the two modalities of PET and MRI (Kashikura et al. 2001). Jack Belliveau from Massachusetts General Hospital (MGH), who at that time was a front runner in the development of functional MRI measurement sequences, sometimes phoned us at around noon Japan Time (1:00 am Boston Daylight Time) to ask what interesting PET results we had obtained that day (Belliveau et al. 1990). From 1990 to 2000 onward, fMRI experiments were a worldwide trend. Most fMRI research studies using an MRI scanner were performed around midnight because the machines were occupied clinically during the daytime.

8.2.3 The BOLD Signal Is a Gift from Nature

During neural activity, cerebral blood flow increases excessively compared to cerebral oxygen metabolism, resulting in the biological phenomenon of an excess in the concentration of local oxygenated hemoglobin, and the physical phenomenon of a change in the magnetic field due to the difference in magnetic susceptibility between oxygenated and deoxygenated hemoglobin. The BOLD contrast is created by a miracle combination of these biological and physical phenomena. If either of these phenomena were missing, the BOLD signal would not exist. Why Nature gave us such an important gift is a great mystery.

If cerebral blood flow and cerebral oxygen metabolism changed in parallel during brain activity, or if hemoglobin had the same magnetic susceptibility in the oxygenated and deoxygenated states, there would be no BOLD contrast. If there were no BOLD contrast, it would have taken a long time to develop a functional brain mapping technique of the same power, and most of the vast amount of information obtained with fMRI, which relies on BOLD contrast, would have remained unknown even today. This means that brain science would still be in the dark. Again, the BOLD contrast is a great gift from Nature. It is a coincidental combination of the biological fact of excess cerebral blood flow relative to cerebral oxygen metabolism and the physical fact of the different magnetic susceptibilities of oxygenated and deoxygenated hemoglobin. The author has nothing but admiration for Seiji Ogawa's keen insight in understanding that this coincidental combination is the source of BOLD contrast and that it could be used for brain functional research.

8.2.4 Paradigm Shift Caused by BOLD Contrast

In the past, when brain functional mapping was performed using oxygen-15 water and PET, the number of experiments was limited to minimize the radiation exposure of the subject, and much energy was devoted to designing the optimal paradigm for mapping the activation focus in the brain with the minimum number of tasks (= number of PET measurements). In addition, there have been many trials and errors to ensure that the observed focus accurately represents the intended activation focus. This led many PET centers to collaborate with neuropsychologists who specialize in neural circuits and brain networks. An early paradigm for activation measurements was the search for activation foci in the somatosensory areas. The research community began by confirming the primary sensory and motor areas, as labeled on Brodmann's map of the cerebral cortex, and the somatosensory and audiovisual areas in healthy adults, as shown in Penfield's Homunculus. The author performed many similar measurements to help validate the methodology for measuring brain activity (Fig. 8.11). The number of measurements that could be made with oxygen-15 water PET mapping was limited due to the radiation exposure of the subjects, so the research mainly concentrated on exploring single (or one-dimensional) activation foci.

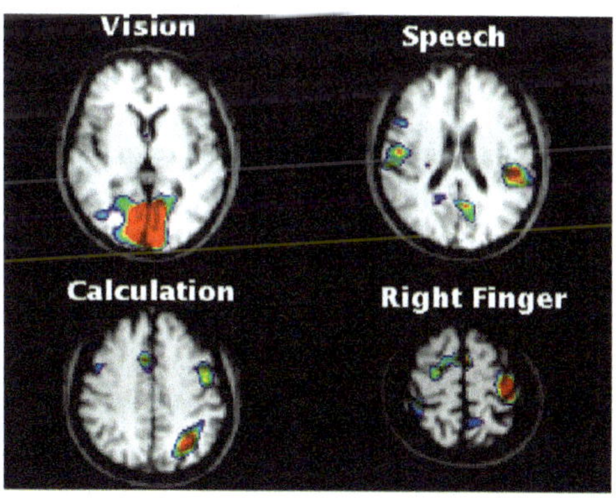

Fig. 8.11 An early attempt by the author at functional mapping using oxygen-15 water PET: Mapping brain function using cerebral blood flow and PET was robust. The activated area was identified from a single performance of the task. (Left and right of picture is right and left-side of the subject, respectively)

Column: Penfield's Homunculus

In the 1930s, Penfield, a neurosurgeon in Montreal, Canada, and colleagues opened the skull of an awake epilepsy patient and electrically stimulated the brain surface of the cortex for therapeutic purposes. They found that it brought back memories of the past to the patient. When the brain was stimulated with electrodes, Penfield recorded positions in the brain where the patient noted sensations of motion, audible sound, and visible light. He then drew pictures of the torso and various limbs along the cortex on a coronal section of the brain. A dwarf-like doll with abnormally large fingers and face and a small torso came into view. Body parts with a high density of peripheral nerves, such as the fingers and face, occupy a large area of the cerebral cortex, while the trunk, with its relatively sparse peripheral nerves, occupies only a small area of the cerebral cortex. This is known as Penfield's Homunculus. These were the first observations that specifically illustrated the functional localization of the brain (Fig. 8.12) (Penfield 1947). At that time, the functional localization of brain functions was confirmed by invasive methods, such as directly touching the brain surface with electrodes through a craniotomy. I do not think Penfield could have imagined that the time would come when MRI could be used to measure brain functions in a noninvasive manner without craniotomy.

However, as MRI scanners capable of performing fMRI with EPI have become more widespread, BOLD contrast measurements are possible at many research and medical facilities. The field has expanded to include both basic and clinical researchers, and a paradigm shift in brain function mapping research has occurred. fMRI-BOLD measurement has made it easier to measure a wide

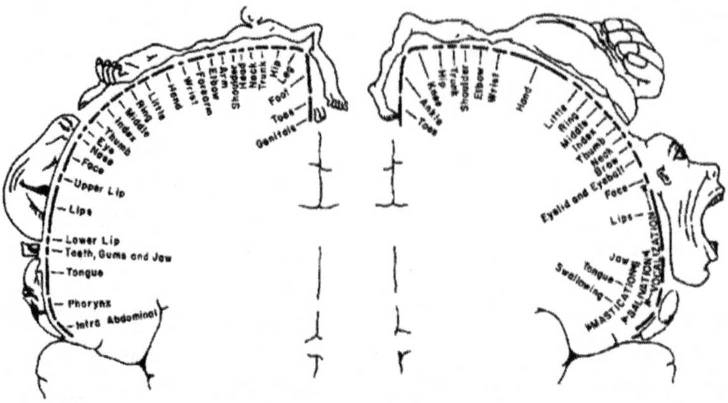

Fig. 8.12 Penfield's Homunculus (dwarf): Penfield measured EEG with electrodes in epilepsy patients, and on a coronal section of the brain he drew body parts in proportion to the cortex area corresponding to each body part. He examined both the somatosensory areas (left) and the motor areas (right). (Penfield W. Proc R SocLondB BiolSci. 1947 134: 329–47)

range of neuroscience related issues. Even researchers in the field of psychology have joined in the search for a functional brain network that links activation foci. fMRI-BOLD has also made it possible to conduct high-precision animal experiments to verify aspects of microcirculation physiology, which cannot be done in humans. Furthermore, with the improved performance of MRI systems, rather than identifying two-dimensional networks with specific activation tasks, it has become possible to perform fMRI measurements that cover the entire brain, enabling the search for three-dimensional networks across the whole brain. The ability to freely measure BOLD imaging over the whole brain has enabled us to detect new brain functional networks, such as the default mode network (DMN), from 3D BOLD measurements made without the subject performing a task. This new direction in brain research, called resting-state fMRI (rsfMRI), is described further below (Raichle 2015) (see Sect. 5.4.2).

8.2.5 Dynamic Range of the BOLD Signal

It should be noted that the dynamic range of the BOLD signal is narrower than that of cerebral blood flow. For example, the authors reported using PET that when the baseline cerebral blood flow was shifted during hypercapnia (e.g., by CO_2 inhalation) or hypocapnia (e.g., by hyperventilation), the rate of change of activated cerebral blood flow increased or decreased in proportion to the baseline cerebral blood flow (Shimosegawa et al. 1995) (see Sect. 8.1.1). However, when the BOLD signal change is measured under the same physiological conditions, the change in the BOLD signal decreases as the baseline increases in hypercapnia, a tendency known as the "ceiling effect" of the BOLD contrast. This phenomenon does not occur with PET cerebral blood flow, which has a wide dynamic range. In other words, the BOLD signal, which has a narrower dynamic range than cerebral blood flow, is not guaranteed to have a linear relationship with neural activity over a wide measurement range. In addition, the baseline, which corresponds to the resting state of MRI-BOLD, can easily drift and should be carefully analyzed. For example, isoflurane anesthesia, which is widely used in animal experiments, has a vasodilation effect so that the cerebral blood flow during resting is relatively high, and a ceiling effect is likely to occur, requiring careful analysis to avoid misinterpretation.

8.2.6 Warning: BOLD Is a Secondary Hemodynamic Signal

The author has had some concerns lately. It is easy for young researchers to forget that the BOLD signal is a secondary hemodynamic signal and mistakenly think that the BOLD signal is a direct expression of neural activity itself. Of course, when a technique is used routinely it is natural for the user to think of it as a black box without understanding the underlying principles. Recently, BOLD-fMRI measurements have been expanded to include not only healthy adults but also dementia and aged patients. However, BOLD contrast is based on the assumption that healthy cerebral blood flow is guaranteed. If an aging person has some cerebrovascular deficit or healthy cerebral blood flow metabolism is not guaranteed, the interpretation of the BOLD signal may be compromised in situations where there is a reduced circulation reserve. Since the discovery of the BOLD contrast, the author is afraid that psychologists, neuropsychiatrists, and naive young researchers, who are newcomers to the field, might interpret the data without a proper understanding of the physiological basis of the BOLD signal, which is secondary to the hemodynamics of cerebral blood flow.

8.3 The Missing Link Between Neural Activity and Cerebral Blood Flow

Circumstantial evidence has emerged that cerebral blood flow is a proxy for neuronal activity. However, the physiological mechanism that links macroscopic observations with neuronal activity and the microscopic phenomenon regulating cerebral microcirculation is not yet understood. The following is what has been confirmed thus far, temporally, spatially, and quantitatively, in *in vivo* experiments.

8.3.1 Does Cerebral Blood Flow Represent Neural Activity?

Based on his own measurements, Lassen made the metabolic substrate feedback theory in the 1970s (See Sect. 8.1.1). The theory proposed that regional cerebral blood flow increased to replenish the metabolic substrates required for regional brain activity, and this theory was widely accepted. However, Fox and Raichle's own PET measurements overturned Lassen's theory by showing that the increase in cerebral oxygen metabolism was reduced compared to the increase in cerebral blood flow. Later, when Lassen learned that cerebral blood flow was excessively increased in comparison to oxygen metabolism, the author remembers that Lassen asked me in a riddle-like way, "Do you know why?" Lassen seemed as if he had anticipated Fox and Raichle's finding and had been satisfied that cerebral blood flow appeared to be the most sensitive marker of brain function. However, in retrospect, there is no reason for cerebral blood flow to be a direct index of neural activity. Rather, the cerebral oxygen metabolic rate, which is a direct indicator of energy consumption during neural activity, is more appropriate. The reality is, however, changes in cerebral oxygen metabolism during activation are reduced, and the signal-to-noise ratio of the measured quantity is low, while changes in cerebral blood flow are large, and the signal-to-noise ratio of the measured quantity is high, making cerebral blood flow a more appropriate quantity to measure on behalf of neural activity. Brain imaging based on changes in cerebral blood flow has expanded its application dramatically with the realization of fMRI-BOLD, a noninvasive measurement technique with an even higher signal-to-noise ratio. Cerebral blood flow, which is merely part of the infrastructure that supports brain tissue, has "transformed" into the most faithful representation of brain function that we have (Lassen and Kanno 1997). It is as if cerebral blood flow plays the role of an amplifier that directly reflects neural activity.

8.3.2 Comparison of Cerebral Blood Flow and Neural Activity

There is a mouse experiment in which neural activity and cerebral blood flow were directly compared. Using genetically modified mice expressing G-CaMP, which allows Ca ions to be visualized during neuronal activity, the experimenters simultaneously measured LSFI (= cerebral blood flow) and Ca firing (= neural activity) of cerebral neurons when peripheral nerves were stimulated. Since changes in cerebral blood flow are much slower than those in neural activity, the map of neural activity was smoothed spatially and temporally before it was compared with the map of cerebral blood flow. The two maps showed excellent agreement (Ohki et al. 2005; Matsui et al. 2019).

Another basic study confirmed the linearity between neuronal activity and cerebral blood flow response. Logothetis et al. performed precise simultaneous measurements of neuronal firing, local field potential (LFP) and BOLD contrast in monkeys and confirmed that BOLD contrast is strictly correlated with LFP, which corresponds to neuronal activity rather than neuronal firing (Logothetis et al. 2001). In animal experiments with rats, the author and his colleague confirmed the linear relationship between cerebral blood flow, measured with a laser Doppler flowmeter (LDF), and LFP during electrical stimulation of the hind paw (See Sect. 8.1.4, Fig. 8.7) (Matsuura and Kanno 2001).

8.3.3 Spatiotemporal Distribution of Cerebral Blood Flow

Improvements in the spatial and temporal resolution of the BOLD signal have led to advances in understanding the mechanism regulating cerebral microcirculation. Simultaneous BOLD signal and two-photon excitation microscopy measurements during electrophysiological stimulation, which are not possible in humans, allow the spatiotemporal relationship between neural activity and cerebral blood flow to be precisely confirmed. Using a spatially and temporally precise experimental paradigm, simultaneous measurements of cerebral blood flow, cerebral blood vessel diameter, and neural activity (LFP) were performed, and the results showed that the region of neural activation and the region of change in cerebral blood flow (BOLD signal) are precisely and faithfully matched in time and space (Logothetis et al. 2001). In addition, improvements in MRI performance and sequence technology have improved imaging resolution, which in turn has led to more reliable confirmation of neural activation in functional anatomy. In small animals such as mice and rats, a spatial resolution of less than 10 μm is possible, and imaging at a temporal resolution as fast as tens of milliseconds has also been performed (Logothetis et al. 2001). In human BOLD signals, a spatial resolution of less than 1 mm and temporal resolution of less than 1 s have become possible (Cheng et al. 2001).

8.3.4 Additivity of Cerebral Blood Flow

Petersen quantitatively measured cerebral blood flow changes during language tasks such as language recall, inscription, reading aloud, and reading silently with oxygen-15-labeled water PET (Petersen et al. 1988). He psychologically analyzed the role of each region of the cerebral cortex involved in language processing based on the subtraction of cerebral blood flow for each task and reported that each region involved in language processing has a cerebral blood flow response according to an additive law. In other words, we can quantitatively confirm that when a certain brain region is active in both task A and task B, the increase in cerebral blood flow in that region is the sum of the increases in cerebral blood flow for task A and task B individually. Based on this additivity, a paradigm for the activation of higher brain functions has been designed. Additivity is also an important principle that is a prerequisite for higher brain function mapping using BOLD signals.

8.4 Resting-State: From Rubbish to Treasure

Up until approximately 2000, functional brain mapping research was dominated by the measurement of cerebral blood flow with oxygen-15-labeled water PET. In the new century after 2000, a new era has begun with researchers from various fields entering BOLD-MRI. However, the unavoidable problem of the resting-state remained.

8.4.1 The Resting-State Was Thought to Be Rubbish

In the early days of brain function mapping, an activation task paradigm was designed to fit the working hypothesis, and the corresponding activation focus was obtained by subtracting the resting-state and the activation state images. The problem here was "What is the resting-state?" In general, the resting-state can be described as a state of "idling," but it is very difficult to maintain a state of "idling" in the brain. Researchers measuring brain activation tasks have always had trouble with this. Most likely, the "state of being idle" varies from subject to subject. In the era of PET, to reduce radiation exposure, it was very important to design the activation task paradigm so that the activation focus could be efficiently extracted from a limited number of measurements. At the same time, we were always worried about what the "resting-state," or control (= reference), would be for the activation task. Raichle worried about this and would always ask at conferences, "What is a resting state?" This problem was a headache for Raichle. It was a particularly vexing question when mapping came to issues of higher brain function. The simplest approach to statistical processing was to consider the "fluctuations" in the "resting state" within and between individuals as "noise" and therefore averaging would statistically cancel out the variability between individuals. For this reason, many subjects were mobilized, many trials were repeated, and the average of these trials was defined as the "resting-state". However, this statistical average cancels out any variation in the unique "resting-state" of an individual as "noise". Another possible problem with this simple statistical processing approach is that it may obscure negative signals during task execution in regions outside the activation focus as "noise". Normally, the resting-state is subtracted from the activation state, and the background levels other than the positive signal from the activation task are ignored as resting-state fluctuations ("noise"). In this way, in the analysis to detect the activation focus for the task, regions other than the activation focus were eliminated as "noise" by the rather violent additive averaging. However, as described in the next section, the fluctuations in the resting-state of individual subjects and the background level of the negative signal might hide important information.

8.4.2 The New Interpretation for the Resting-State

As is well known, even though the brain contributes only 2% of the body's weight, the brain consumes 20% of the oxygen consumed by the entire body in the resting-state. In other words, the brain consumes 10 times more energy per unit weight than any other tissue or organ in the body. However, it seems that 60–80% of this cerebral oxygen metabolism is used for communication between neurons and the maintenance of the neurons themselves, and only 20–40% is consumed by neural activity during a task (Raichle 2006). In the cerebral cortex, more energy is used for the endogenous activity of the neurons themselves than expected. Raichle and colleagues noticed this when they performed resting-state fMRI (rsfMRI), rather than conventional fMRI, which explores the brain's active areas in response to a task. They found that in regions where the BOLD signal had previously been elevated during an fMRI task, the BOLD signal decreased in the resting state. Conversely, in regions where the BOLD signal had decreased during an activation task it increased in the resting-state (Raichle 2015). This led Raichle to hypothesize that regions with elevated BOLD signals on rsfMRI had increased cerebral blood flow in the resting

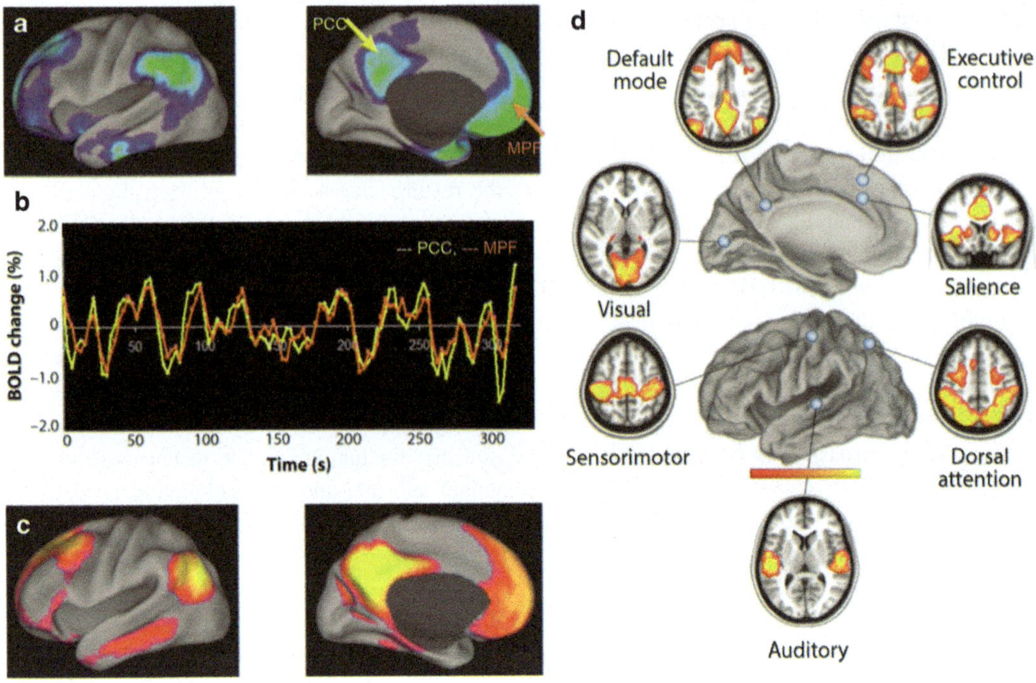

Fig. 8.13 Default mode network: (**a**) The hypoactive parts of the posterior cingulate cortex (PCC) and medial prefrontal cortex (MPF). (**b**) BOLD response in the PCC and MPF during task execution. (**c**) Functionally correlated areas according to the BOLD response. (**d**) Resting-state patterns of functional connectivity. (Raichle ME Ann Rev. Neurosci 2015)

state. Furthermore, careful analysis of these regions showed that the BOLD signals in these regions were highly correlated with each other. They named these regions the default mode network (DMN) and suggested that it is the command center of the brain. When the rsfMRI BOLD signals were correlated with core brain regions such as the visual and auditory cortices, strong correlations were found, and the networks related to each region were identified (Fig. 8.13). The physiological significance of the networks between brain regions obtained from the rsfMRI BOLD measurements is still not fully understood, but the energy spent in the "default mode," or baseline, activity of these brains is estimated to be 20 times more than the energy that the brain uses for conscious task activity. The BOLD signal obtained with rsfMRI can be interpreted as an indicator of unconscious brain energy consumption. It has been hypothesized that the role of the networks is to synchronize various neural activities in the brain. It is also thought to be an idling state in preparation for possible future events. The "resting-state" has changed from being trash to being a rich new mine for neuro-/physio-scientific treasure.

References

Attwell D, Iadecola C. The neural basis of functional brain imaging signals. Trends Neurosci. 2002;25(12):621–5. Review.

Attwell D, Buchan AM, et al. Glial and neuronal control of brain blood flow. Nature. 2010;468:232–43. Review.

Belliveau JW, Rosen BR, et al. Functional cerebral imaging by susceptibility-contrast NMR. Magn Reson Med. 1990;14(3):538–46.

Cheng K, et al. Human ocular dominance columns as revealed by high-field functional magnetic resonance imaging. Neuron. 2001;32(2):359–74.

Fox PT, Raichle ME. Focal physiological uncoupling of cerebral blood flow and oxidative metabolism during somatosensory stimulation in human subjects. Proc Natl Acad Sci USA. 1986;83(4):1140–4.

Fox PT, Raichle ME, et al. Nonoxidative glucose consumption during focal physiologic neural activity. Science. 1988;241(4864):462–4.

Frackowiak RSJ, Magistretti PJ, et al. editors. Neuroenergetics: relavance for functional brain imaging. HFSP Workshop XI, Strasbourg, 2001.

Goyal MS, Hawrylycz M, et al. Aerobic glycolysis in the human brain is associated with development and neotenous gene expression. Cell Metab. 2014;19(1):49–57.

Iadecola C. The neurovascular unit coming of age: a journey through neurovascular coupling in health and disease. Neuron. 2017;96(1):17–42.

Kashikura K, Kershaw J, et al. Temporal characteristics of event-related BOLD response and visual-evoked potentials from checkerboard stimulation of human V1: a comparison between different control features. Magn Reson Med. 2001;45(2):212–6.

Lassen NA, Kanno I. The metabolic and hemodynamic events secondary to functional activation—notes from a workshop held in Akita, Japan. Magn Reson Med. 1997;38(4):521–3. Review.

Lassen NA et al., editors. Brain work and mental activity: quantitative studies with radioactive tracers (brain work II): Proceedings for the Alfred Benzon symposium XXXI, Copenhagen, 12–16 August, 1990.

Logothetis NK, Pauls J, et al. Neurophysiological investigation of the basis of the fMRI signal. Nature. 2001;412(6843):150–7.

Matsui T, Murakami T, Ohki K. Neuronal origin of the temporal dynamics of spontaneous BOLD activity correlation. Cereb Cortex. 2019;29:1496–508.

Matsuura T, Kanno I. Quantitative and temporal relationship between local cerebral blood flow and neuronal activation induced by somatosensory stimulation in rats. Neurosci Res. 2001;40(3):281–90.

Matsuura T, Kashikura K, Kanno I. Hemodynamics of local cerebral blood flow induced by somatosensory stimulation under normoxia and hyperoxia in rats. Comp Biochem Physiol A Mol Integr Physiol. 2001;129(2–3):363–72.

Ogawa S, Lee TM, et al. Brain magnetic resonance imaging with contrast dependent on blood oxygenation. Proc Natl Acad Sci USA. 1990;87(24):9868–72.

Ohki K, Chung S, et al. Functional imaging with cellular resolution reveals precise microarchitecture in visual cortex. Nature. 2005;433(7026):597–603.

Penfield W. Some observations on the cerebral cortex of man. Proc R Soc Lond B Biol Sci. 1947;134(876):329–47.

Petersen SE, Fox PT, et al. Positron emission tomographic studies of the cortical anatomy of single-word processing. Nature. 1988;331(6157):585–9.

Phelps ME, Mazziotta JC. Positron emission tomography: human brain function and biochemistry. Science. 1985;228(4701):799–809. Review.

Raichle ME. Neuroscience. The brain's dark energy. Science. 2006;314(5803):1249–50.

Raichle ME. Behind the scenes of functional brain imaging: a historical and physiological perspective. Proc Natl Acad Sci USA. 1998;95(3):765–72. Review.

Raichle ME. The brain's default mode network. Annu Rev Neurosci. 2015;38:433–47.

Raichle ME, Mintun MA. Brain work and brain imaging. Annu Rev Neurosci. 2006;29:449–76. Review.

Shimosegawa E, Kanno I, et al. Photic stimulation study of changing the arterial partial pressure level of carbon dioxide. J Cereb Blood Flow Metab. 1995;15(1):111–4.

Zheng Y, Pan Y, et al. A dynamic model of neurovascular coupling: implications for blood vessel dilation and constriction. NeuroImage. 2010;52(3):1135–47.

Epilogue

9

Abstract

Near the end of the last century the brain was seen as the last frontier of the human body, and there was a strong motivation to elucidate the mechanisms of the brain because it would advance information technology and science. However, this enthusiasm has gradually faded away. Rapid progress in other branches of the biological and medical sciences, such as genetics, immunology, regenerative medicine, has attracted the attention of young scientists. Nevertheless, there are still many unknowns in the relationship between cerebral blood flow and brain activity. What we understand now is only the tip of an iceberg floating in the ocean. The author hopes that this book will help inspire young scientists to reassess cerebral blood flow as a tool for mining the vast unknown treasure of the brain.

Keywords

Last frontier · Missing link · Phenotype · Genetics · Regenerative medicine · Immunology

9.1 Ebbing of Enthusiasm

At the beginning of the twenty-first century, the new century was called the "Century of the Brain." The brain was touted as the last frontier of the human body. In fact, the range of brain imaging has expanded beyond neuroscience. It has become a fundamental technology in many branches of clinical medicine, such as neurosurgery and neurology, and an essential tool for many areas of clinical research, including rehabilitation and psychiatry. Furthermore, it has had a great impact on the fields of neurosociology and neuroeconomics, which are beyond the framework of basic physiology. However, in recent years the enthusiasm for fundamental neuroscience has been ebbing a little, and if anything, young researchers have become more interested in technologies such as VR (virtual reality), IoT (Internet of things, network technology), AI (artificial intelligence), machine learning, big data analysis, and even genetic engineering, which are all fields that have been progressing with extreme speed. Instead of trying to learn from the brain, we seem to be focusing on creating a new reality with the latest exciting technologies. Rather, the brain is seen as nothing more than a mass of low-tech networks, and it is no longer an attractive subject of research that we can learn from. For today's young scientists, the new fields of genetics, regenerative medicine, immunology, and so on seem to be more spectacular.

9.2 Phenotype of Brain Activity

However, there is still an unsolved mystery surrounding cerebral blood flow, which supports the brain by supplying it with nutrients. Why is it that cerebral blood flow, which is mere infrastructure, is the most reliable and highly accurate method for

I. Kanno, *What Cerebral Blood Flow Tells Us*, https://doi.org/10.1007/978-981-97-6382-5_9

investigating brain activity? Furthermore, from a layman's point of view, it is the actual energy consumption, or the cerebral oxygen metabolic rate, associated with neural activity that should most faithfully reflect brain function. Ironically, however, cerebral blood flow (or BOLD signal), which is supposed to be just infrastructure, is the most sensitive biomarker of brain activity. Is this inevitable, coincidental, or is there an unknown reason behind it? It seems to the author that Nature has its reasons, but for now it is keeping them to itself.

Column: Is Cerebral Blood Flow Infrastructural?

For a long time, the author measured cerebral blood flow at a stroke center in a cold, snowy country. I learned there that if cerebral blood flow is blocked due to stroke, the brain immediately stops functioning, and so cerebral blood flow is the almighty power that controls brain function. One day, I had an opportunity to attend a conference with a neuroscientist at a cross-disciplinary forum. During our conversation I emphasized that cerebral blood flow controls brain function and can be used to map the brain because I had started brain function mapping with PET. However, the neuroscientist did not agree, saying, "No matter how versatile it is, cerebral blood flow is an infrastructure." In other words, "cerebral blood flow only supports the brain, but it is actively incompetent." Later, when the author attended the Society for Neuroscience meeting, the world's largest neuroscience conference with tens of thousands of attendees, not even 10% of the presentations were on cerebral blood vessels, cerebral blood flow, or cerebral ischemia. This made me realize how minor cerebral blood flow is in neuroscience and that the world view of the brain based on cerebral blood flow is like that of a "frog in a well". However, I still wonder whether cerebral blood flow is just infrastructure.

9.3 Remaining Questions

Even more curious to the author is that in quantitative PET measurements it is not the absolute value of cerebral blood flow that reflects neural activity but the relative value after normalization by the whole brain average cerebral blood flow. This is seemingly contradictory to the quantitative relationship between neural activity and cerebral blood flow discussed throughout this book. The phenomenon of the relative value being meaningful has also been confirmed by localized measurements of neural activity (with EEG) and cerebral blood flow. It is not known why, but it can be said that, rather than absolute cerebral blood flow, it is relative cerebral blood flow, which reflects neural activity, that has a linear relationship with the average value of the whole brain. There is still an unknown missing link between neuronal activity and localized blood flow changes in capillaries and small arteries.

9.4 Tip of the Iceberg

Recent advances in *in vivo* (live) microscopic imaging of the brain have made it possible to see the functions of individual cells in the brain. The functions of the individual constituent cells of the NVU have been visualized through advances in genetic manipulation and two-photon excitation microscopy, and the integration of this information should increase understanding of the brain as a whole. Since the 1990s, Raichle has often used slides of icebergs floating on the ocean surface in his lectures while saying that only about the 10% that is above the surface is known, and almost nothing is known about the remaining 90% (Fig. 9.1). Even though we are trying to explore the unknown 90%, the author is afraid that the mystery beneath the ocean surface will not be easily revealed, but rather, the mystery will increase.

Fig. 9.1 An iceberg floating on the ocean: Only 10% of the brain's activity is visible to us above sea level. The other 90% of the brain activity is below sea level and is largely unknown and yet to be explored

9.5 Unknown Treasures

It is not known what kind of breakthroughs will occur in the future and how the technology for measuring brain signals will be developed. Perhaps the time will come when it will be possible to directly measure all neural firing events as neural activity in a noninvasive manner. If this happens, massive data sets containing the activity of approximately one hundred or several hundred billion neurons and glial cells will be measured in real time. Even with such information there will be many questions remaining, like how to construct an overall picture of the brain, how individual nerves relate to other nerves in the entire brain, and how to integrate large amounts of measured data into a single model. Will it be possible to solve the mystery? The brain is a trove of vast unknown treasures. This author hopes that young scientists will continue to dig into this mountain of treasure to uncover new brain functions, mental functions, and rich mineral veins between the brain and mind.

The manufacturer's authorised representative in the EU is Springer Nature Customer Service Centre GmbH, Europaplatz 3, 69115 Heidelberg, Germany. If you have any concerns regarding our products, please contact ProductSafety@springernature.com

Printed and bound by CPI Group (UK) Ltd, Croydon, CR0 4YY
30/09/2026
02225265-0019